795#0

D0160495

# TOWARD A NATIONAL URBAN POLICY

# TOWARD
# A NATIONAL
# URBAN POLICY

*Edited by* DANIEL P. MOYNIHAN

BASIC BOOKS, INC., PUBLISHERS

*New York*                    *London*

HT
123
.T66

SECOND PRINTING

© 1970 by Basic Books, Inc.
Library of Congress Catalog Card Number: 79-103092
SBN 465-08627-6
Manufactured in the United States of America

# The Authors

ROBERT A. DENTLER, Professor of Sociology and Education at Columbia University Teachers College, is Director of the Center for Urban Education. Among his published works are *The Politics of Urban Renewal, Big City Dropouts and Illiterates,* and *American Community Problems.*

EDWARD P. EICHLER, currently Vice-President of The Klingbeil Company, has served as Special Advisor to the Department of Housing and Urban Development (1966) and as U.S. Representative to the U.N. Conference on Urban Development in the Soviet Union (1964).

MARION B. FOLSOM recently retired as Director and former Treasurer of Eastman Kodak. He served as Under Secretary of the Treasury, 1953–1955; and as Secretary of Health, Education, and Welfare, 1955–1958. From 1962 to 1967 he headed the National Advisory Commission on Community Health Services.

NATHAN GLAZER is Professor of Education and Social Structure at the Harvard University Graduate School of Education. He has served as Urban Sociologist for the Housing and Home Finance Agency, and his books include *Studies in Housing and Minority Groups* (co-edited with Davis McEntire) and *Beyond the Melting Pot* (with Daniel P. Moynihan).

ANDREW M. GREELEY is Program Director of the National Opinion Research Center and Lecturer in Sociology at the University of Chicago. His published works include *The Education of Catholic Americans* (with Peter H. Rossi); *Why Can't They Be Like Us?: Ethnic Conflict in America;* and *Religion in the Year 2000.*

SCOTT A. GREER is Professor of Sociology and Political Science at Northwestern University. His latest published works include the following: *The New Urbanization* (edited with Dennis McElrath, David W. Minar, and Peter Orleans); *The Logic of Social Inquiry;* and *The Concept of Community* (with David W. Minar).

ROBERT GUTMAN, a specialist in the sociology of architecture,

v

29225

housing and building, is Professor of Sociology at Rutgers University. Most recently he has published *Urban Sociology: A Bibliography* and *Site Planning and Social Organization.*
PHILIP M. HAUSER's published works include *The Study of Urbanization; Urbanization in Latin America;* and *Urbanization in Asia and the Far East.* He is Professor of Sociology and Director of the Population Research Center, University of Chicago.
GLENN R. HILST, the Executive Vice-President for Research of the Travelers Research Corporation, lectures at the Yale University School of Medicine. His works include numerous technical articles on environmental effects and their control.
JOHN F. KAIN, a specialist in urban problems, is co-author of *The Urban Transportation Problem* and Professor of Economics at Harvard University.
RICHARD C. LEE, Mayor of New Haven, Connecticut, is a member of the Advisory Committee and past President of the United States Conference of Mayors. During the 1960 presidential campaign, he served as advisor on urban affairs to the late John F. Kennedy.
JOHN R. MEYER is Professor of Economics at Yale University and President of the National Bureau of Economic Research in New York City. Dr. Meyer's work includes *The Urban Transportation Problem; Technology and Urban Transportation;* also *Investment Decisions, Economic Forecasting and Public Policy.*
MARTIN MEYERSON is President of the State University of New York at Buffalo. He was formerly Dean of the College of Environmental Design and Acting Chancellor, University of California at Berkeley. Prior to that he served as Williams Professor of City Planning and Urban Research, Harvard University, and Director of the M.I.T.-Harvard University Joint Center for Urban Studies. His most recent publication (with Edward C. Banfield) is *Boston: The Job Ahead.*
DANIEL P. MOYNIHAN is currently Counsellor to the President. On leave as Professor of Education and Urban Politics at Harvard University, he is editor of a volume in the Basic Books Perspectives on Poverty series *On Understanding Poverty.* Dr. Moynihan is the author of *Maximum Feasible Misunderstanding* and *Beyond the Melting Pot* (with Nathan Glazer).

BERNARD NORWITCH is a consultant to the National Affairs Division of the Ford Foundation. He was a Vice-President of Reston, Va., Inc., and was a U. S. Senate aide. He has written for several journals.

THOMAS F. PETTIGREW, Professor of Social Psychology at Harvard University, has written *A Profile of the Negro American* and *Racially Separate or Together?*

FRANCINE F. RABINOVITZ is the author of *City Politics and Planning*. She is Assistant Professor of Political Science at the University of California at Los Angeles.

LEE RAINWATER is Professor of Sociology at Harvard University. Among his books are *Behind Ghetto Walls: Black Families in a Federal Slum* and *The Moynihan Report and the Politics of Controversy*.

MARTIN REIN, co-author with Peter Marris of *Dilemmas of Social Reform: Poverty and Community Action in the United States* and author of *Introduction to Social Policy*, is Graduate Professor of Social Policy at Bryn Mawr College.

LLOYD RODWIN serves as Director of the Special Program for Urban and Regional Studies of Developing Areas and as Professor of Urban Studies and Planning at M.I.T. The editor of *The Future Metropolis* and *Planning Urban Growth and Regional Development*, his latest book is *Nations and Cities*.

CHARLES TILLY, Professor of Sociology and History at the University of Michigan, teaches courses on cities, urbanization, and political change. In *The Vendee, Race and Residence in Wilmington,* and other works, he has explored violent protest, political conflict, urbanization, migration, and urban structure in Europe and America.

JOHN F. C. TURNER is a Lecturer at M.I.T. and Consultant to the United Nations and the Organization of American States. The author of a number of published papers on low-income housing in developing areas, he is completing a book on urban settlement, squatters, and social change which is being prepared under the auspices of the Joint Center for Urban Studies of M.I.T. and Harvard University.

WOLF VON ECKARDT is the architecture critic of *The Washington Post*. His books include *Eric Mendelsohn* and *A Place to Live: The Crisis of the Cities*.

JAMES Q. WILSON is Chairman of the Department of Government and Professor of Government at Harvard University where he has taught since 1961. From 1963 to 1966, he was Director of the Joint Center for Urban Studies of M.I.T. and Harvard University. He was also a member of the Science Advisory Committee to the President's Commission on Law Enforcement and Administration of Justice. The author of a number of books and articles on urban affairs, his most recent publication is *Varieties of Police Behavior.*

ROBERT WOOD is Chairman of the Department of Political Science at M.I.T. and Director of the Joint Center for Urban Studies of M.I.T. and Harvard. He was Under Secretary of Housing and Urban Development from 1966 to 1969. He is co-author of *Politics and Government in the United States.*

# Preface

I write this brief preface on the morning of the day Americans are landing on the moon. More than a master work of technology, the landing of Eagle is a triumph of national will, the end result of eight years and fifty-six days of effort from the moment John F. Kennedy proposed that we send a man to the moon and return him safely, and do so with the "full speed of freedom." Yet the air of victory is curiously muted. Triumph in space seems only to intensify the concern in academic, intellectual, journalistic, and some political circles, with our seeming inability to get things done on earth, especially with respect to cities.

Concern for the condition of American cities, and the emerging sense that some coherent national approach needs to be made to the problems to be encountered in cities, has come near to being an obsession of our time.

Obsession is not too strong a word; nor is it to be judged an overreaction. The life in American cities has come to be singularly troubled, and it is peculiarly the fate of Americans to see themselves as directly implicated and challenged by that seemingly impersonal development.

If, in the European past, men could properly be seen as having been shaped by their cities—and so regarded themselves—it is here, first on the coastal fringe, thence through the great forest, the vast plain beyond, the mountains, and finally to the farther coast, that cities were shaped by men. They had not emerged from the past: they had been founded in the present, by contemporary men, and not infrequently named for such. Nothing quite like those cities has ever existed: nothing quite so good; also, nothing perhaps quite so awful. Some part of the profligate energy and endless resource that went into building their cities went also into contemplating them, and a measure of this contemplation was tinged with fear. So long as the nation was technically "rural" it could be supposed that fear of the city was in part at least a matter of nostalgia for a simpler past, somewhat structured by a Jeffersonian ideological heritage. ("In the beginning," as the Whites put it, "was the farm.") But by 1920 the census reported that a majority of the American population was in fact urban; half a century later the proportion has reached three quarters. To live in a city (although not necessarily an enormous one) is now normal for the American: a clear majority have known nothing else.

And yet the fear persists, and so also the fascination. It was with both concerns in mind that Mr. Theodore Wertime, then editor of the Voice of America Forum series, conceived the collection of essays of which this volume consists. Looking about for someone in the "presidential government" who might assemble contributors with special interests in the area of public policy, he turned to me. This was in the spring of 1964. Although by that time the federal establishment included a Housing and Home Finance Agency, there was as yet no urban department, certainly no place where urban concerns were focused. The Department of Labor, curiously, was as much such a center as any. Its constituency was almost by definition urban and its thoughts were directed to the city and the enormous strains just then beginning to appear in the Northern urban complexes in the aftermath of the great internal migration of the 1940's and 1950's. I was then Assistant Secretary of Labor for Policy Planning and Research, and so involved with these developments. The proposal was intriguing: to gather in one volume the views of men of disparate views and interests but one central concern, namely the formulation in the course of the 1960's and 1970's of something approaching a deliberate and elaborated national urban policy. Our purpose would be to share such thoughts as we might have with the international audience of the Forum series, and thereafter to publish them for the American reader. It was not the clearest of undertakings, but it seemed worthwhile.

It has, of course, taken longer than it ought to have. I did not finish the revisions of my final essay until December, 1968, and yet a certain symmetry of sequence emerges. Within weeks President-elect Richard Nixon had asked me to return to Washington as Assistant to the President for Urban Affairs. The first executive order of his administration established the Council for Urban Affairs, charging it with the task of assisting the president in the "development of a national urban policy." How successful the enterprise will be, if the notion of success is even applicable here, is matter yet to be known, if indeed knowable. But the effort has begun, as the opening essay, written in the spring of 1969, will attest.

I would like to think that these essays might add something to that effort. The talents of the individual contributors are singular: I do not believe a more distinguished combination of intellect and experience is to be found in this field. I am not so much grateful for their contributions as proud of their company. The rewards of a Forum lecture are modest by any standards, and the demands on the time and creativity of the men who wrote, then, typically, rewrote and expanded the essays of this volume are remorseless. Yet very few of the persons invited to contribute declined. Perhaps these facts should be taken also as a measure of the regard and reputation enjoyed by the Voice of America and the Forum series. May it remain such.

I should like also to express my appreciation to Mr. Wertime for the unfailingly gentle pressure which kept the project moving. Irving Kristol was wise and helpful and patient throughout. In that manner he moves mountains.

The volume simply would not exist, however, were it not for Janet Eckstein, whose superb editorial talents and professional understanding of the subject alone made possible such coherence as the volume may have—as the subject may admit.

DANIEL P. MOYNIHAN

*Derrymore*
*West Davenport, New York*
*July 20, 1969*

# Contents

CONTENTS

# TOWARD A NATIONAL URBAN POLICY

# 1 TOWARD A NATIONAL URBAN POLICY*

**Daniel P. Moynihan**

In the Spring of 1969, President Nixon met in the Cabinet room with the mayors of ten American cities. They were nothing if not a variegated lot, mixing party, religion, race, region in the fine confusion of American politics. They had been chosen to be representative in this respect, and were unrepresentative only in qualities of energy and intelligence that would have set them apart in any company. What was more notable about them, however, was that in the interval between the invitation from the White House and the meeting with the president, four had, in effect, resigned. All but assured of re-election, they had announced they would not run again. The mayor of Detroit who, at the last minute, could not attend, announced *his* resignation in June.

Their decisions were not that uncommon. More and more, for the men charged with governance of our cities great and small, politics has become the art of the impossible. It is not to be wondered that they flee. But we, in a sense, are left behind. And are in trouble. And know it.

At a time of great anxiety—a time which one of the nation's leading news magazines now routinely describes as "the most serious domestic crisis since the Civil War," a time when Richard Rovere, writing of the 1972 elections, adds parenthetically "assuming that democracy in America survives that long"—these personal decisions may seem of small consequence, yet one suspects they are not. All agree that the tumult of the time arises, in essence,

* This chapter was published first by special permission in *The Public Interest* (Fall, 1969), pp. 3–20.

from a crisis of authority. The institutions which shaped conduct and behavior in the past are being challenged, or worse, ignored. It is in the nature of authority, as Robert A. Nisbet continues to remind us, that it is consensual, that it is not coercive. When authority systems collapse they are replaced by power systems, which are coercive. Our vocabulary rather fails us here: the term "authority" is an unloved one, with its connotations of authoritarianism, but there appears to be no substitute. Happily, public opinion is not so dependent on political vocabulary, certainly not on the vocabulary of political science, as some assume. For all the ambiguity of the public rhetoric of the moment the desire of the great mass of our people is clear. They sense the advent of a power-based society and they fear it. They seek peace. They look to the restoration of legitimacy, if not in existing institutions, then in new or modified ones. They look for a lessening of violent confrontations at home, and, in great numbers, for an end to war abroad. Concern for personal safety on the part of city dwellers has become a live political fact, while the reappearance—what, praise God, did we do to bring this upon ourselves?—of a Stalinoid rhetoric of apocalyptic abuse on the left, and its echoes on the right, have created a public atmosphere of anxiety and portent that would seem to have touched us all. It is with every good reason that the nation gropes for some means to weather the storm of unreason that has broken upon us and seems if anything to grow wilder.

It would also seem that Americans at this moment are much preoccupied with the issue of freedom, or rather with new, meaningful ways in which freedom is seen to be expanded or constrained. We are, for example, beginning to evolve some sense of the meaning of group freedom. This comes after a century of preoccupation with individual rights of a kind which were seen as somehow opposed to and even threatened by group identities and anything so dubious in conception as "group rights."

The Civil Rights Act of 1964 was the culmination of the political energies generated by that earlier period. The provisions which forbade employers, universities, governments, or whatever to have any knowledge of the race, religion, or national origin of individuals with which they dealt marked in ways the high-water

4

mark of Social Darwinism in America, and did not long stand unopposed. Indeed, by 1965 the federal government had already, as best one can tell, begun to require ethnic and racial census of its own employees, of federal contractors, and research grant recipients. To do so violated the spirit if not the letter of the Civil Rights Act, with its implicit model of the lone individual locked in equal—and remorseless—competition in the Mancunian marketplace, but very much in harmony with the emerging sense of the 1960's that groups have identities and entitlements as well as do individuals. This view is diffusing rapidly. (In Massachusetts, for example, legislation of the Civil Rights Act period that declared any public school with more than 50 per cent black pupils to be racially "imbalanced" and in consequence illegal, is already being challenged—by precisely those who supported it in the first instance.) If so far these demands have been most in evidence among black Americans, there is not the least reason to doubt that they will now diffuse to other groups, defined in various ways, and that new institutions will arise to respond to this new understanding of the nature of community.

In sum, two tendencies would appear to dominate the period. The sense of general community is eroding, and with it the authority of existing relationships, while simultaneously a powerful quest for specific community is emerging in the form of ever more intensive assertions of racial and ethnic identities. Although this is reported in the media largely in terms of black nationalism, it is just as reasonable to identify emergent attitudes in the "white working class," as part of the same phenomenon. The singular quality of these tendencies is that they are at once complementary and opposed. While the ideas are harmonious, the practices that would seem to support one interest are typically seen as opposing the other. Thus one need not be a moral philosopher or a social psychologist to see that much of the "crisis of the cities" arises from the interaction of these intense new demands, and the relative inability of the urban social system to respond to them.

Rightly or otherwise—and one is no longer sure of this—it is our tradition in such circumstances to look to the condition of government. Social responses to changed social requirements take the form in industrial democracies of changed government policies.

This had led, in the present situation, to a reasonably inventive spate of program proposals of the kind the New Deal more or less began and which flourished most notably in the period between the presidential elections of 1960 and 1968 when the number of domestic programs of the federal government increased from 45 to 435. Understandably, however, there has been a diminution of the confidence with which such proposals were formerly regarded. To say the least, there is a certain nonlinearity in the relationship between the number of categorical aid programs issuing forth from Washington, and the degree of social satisfaction that ensues.

Hence the issue arises as to whether the demands of the time are not to be met in terms of policy, as well as program. It has been said of urban planners that they have been traumatized by the realization that everything relates to everything. But this is so, and the perception of it can provide a powerful analytic tool.

Our problems in the area of social peace and individual or group freedom occur in urban settings. Can it be that our difficulties in coping with these problems originate, in some measure, from the inadequacies of the setting in which they arise? Crime on the streets and campus violence may mark the onset of a native nihilism: but in the first instance they represent nothing more complex than the failure of law enforcement. Black rage and white resistance, so-called Third World separatism, and restricted covenants all may define a collapse in the integuments of the social contract: but, again, in the first instance they represent for the most part simply the failure of urban arrangements to meet the expectations of the urban population in the areas of jobs, schools, housing, transportation, public health, administrative responsiveness, and political flexibility. If all these are related, one to the other, and in combination do not seem to be working well, the question arises whether the society ought not attempt a more coherent response. In a word, ought a national urban crisis be met with something like a national urban policy? Ought not the vast efforts to control the situation of the present, be at least informed by some sense of goals for the future?

The United States does not now have an urban policy. The idea that there might be such a policy is new; so also is the Urban Affairs Council, established by President Nixon on January 23,

1969, as the first official act of his administration, to "advise and assist" with respect to urban affairs, specifically "in the development of a national urban policy, having regard both to immediate and to long-range concerns, and to priorities among them."

The central circumstance, as stated, is that America is an urban nation, and has been one for half a century.

This is not to say that most Americans live in large cities. They do not. In 1960 only 9.8 per cent of the population lived in cities populated by one million or more inhabitants. In fact, 98 per cent of the units of local government have fewer than 50,000 persons. In terms of the 1960 census only somewhat more than a quarter of congressmen represented districts in which a majority of residents lived in central city areas. The 1970 census will show that the majority of Americans in metropolitan areas in fact live in suburbs, while a great many more live in urban settlements of quite modest size. But they are not the less urban for that reason, providing conditions of living and problems of government profoundly different from that of the agricultural, small-town past.

The essentials of the present urban crisis are simple enough to relate. Until about World War II the growth of the city, as Otto Eckstein argues, was "a logical, economic development." At least it was such in the northeastern quadrant of the United States, where most urban troubles are supposed to exist. The political jurisdiction of the city more or less defined the area of intensive economic development which more or less defined the area of intensive settlement. Thereafter economic incentives and social desires combined to produce a fractionating process which made it ever more difficult to collect enough power in any one place to provide the rudiments of effective government. As a result of or as a part of this process, the central area ceased to grow and began to decline. The core began to rot. This most primitive analogue began to suggest to us that in some way life itself was in decline.

Two special circumstances compounded this problem. First, the extraordinary migration of the rural Southern Negro to the northern city. Second, a postwar population explosion (90 million babies were born between 1946 and 1968) which placed immense pressures on municipal services, and drove many whites to the

7

suburbs seeking relief. (Both these influences are now attenuating somewhat, but their effects will be present for at least several decades, and indeed a new baby boom may be in the offing.) As a result the problems of economic stagnation of the central city became desperately exacerbated by those of racial tension. In the course of the 1960's tension turned into open racial strife.

City governments began to respond to the onset of economic obsolescence and social rigidity a generation or more ago, but quickly found their fiscal resources strained near to the limit. State governments became involved, and much the same process ensued. Starting in the postwar period, the federal government itself became increasingly caught up with urban problems. In recent years resources on a fairly considerable scale have flowed from Washington to the cities of the land and will clearly continue.

However, in the evolution of a national urban policy, more is involved than merely the question of national goals and the provision of resources with which to attain them. Too many programs have produced too few results simply to accept a more or less straightforward extrapolation of past and present practices into an oversized but familiar future.

The question of method has become as salient as that of goals themselves. As yet the federal government, no more than state or local government, has not found an effective incentive system—comparable to profit in private enterprise, prestige in intellectual activity, rank in military organization—whereby to shape the forces at work in urban areas in such a way that urban goals—whatever they may be—are in fact attained. This search for incentives, and the realization that present procedures such as categorical grant-in-aid programs do not seem to provide sufficiently powerful ones, must accompany and suffuse the effort to establish goals as such. We must seek not just policy, but policy allied to a vigorous strategy for obtaining results from it.

Finally, the federal establishment must develop a much heightened sensitivity to its "hidden" urban policies. There is hardly a department or agency of the national government whose programs do not in some way have important consequences for the life of cities, and those who live in them. Frequently—one is tempted to say normally!—the political appointees and career executives concerned do not see themselves as involved with, much less responsible for the urban consequences of their programs and policies. They are, to their minds, simply building highways, guaranteeing mortgages, advancing agriculture, or whatever. No one has made clear to them that they are simultaneously redistributing employment opportunities, segregating neighborhoods, or desegregating them, depopulating the countryside and filling up the slums, and so forth: all of these things are second and third order consequences of nominally unrelated programs. Already this institutional naivete has become cause for suspicion; in the future it simply must not be tolerated. Indeed, in the future, a primary mark of competence in a federal official should be the ability to see the interconnections between programs immediately at hand, and the urban problems that pervade the larger society.

## THE FUNDAMENTS OF URBAN POLICY

It having long been established that with respect to general codes of behavior eleven precepts are too many, and nine too few, ten points of urban policy may be set forth, scaled roughly to correspond to a combined measure of urgency and importance.

1. The poverty and social isolation of minority groups in central cities is the single most serious problem of the American city today. It must be attacked with urgency, with a greater commitment of resources than has heretofore been the case, and with programs designed especially for this purpose.

The 1960's have seen enormous economic advances among minority groups, especially Negroes. Outside the South, 37 per cent of Negro families earn $8,000 per year or more, that being approximately the national median income. In cities in the largest metropolitan areas, 20 per cent of Negro families in 1967 reported family incomes of $10,000 or over. The earnings of young married black couples are approaching parity with whites.

Nonetheless, certain forms of social disorganization and dependency appear to be increasing among the urban poor. Recently,

9

Conrad Taueber, Associate Director of the Bureau of the Census, reported that in the largest metropolitan areas—those with one million or more inhabitants—"the number of black families with a woman as head increased by 83 per cent since 1960; the number of black families with a man as head increased by only 15 per cent during the same period." Disorganization, isolation, and discrimination seemingly have led to violence, and this violence has in turn been increasingly politicized by those seeking a "confrontation" with "white" society. Urban policy must have as its first goal the transformation of the urban lower class into a stable community based on dependable and adequate income flows, social equality, and social mobility. Efforts to improve the conditions of life in the present caste-created slums must never take precedence over efforts to enable the slum population to disperse throughout the metropolitan areas involved. Urban policy accepts the reality of ethnic neighborhoods based on choice, but asserts that the active intervention of government is called for to enable free choice to include integrated living as the normal option.

It is impossible to comprehend the situation of the black urban poor without first seeing that they have experienced not merely a major migration in the past generation, but also that they now live in a state almost of demographic siege as a result of population growth. The dependency ratio, in terms of children per thousand adult males, for blacks is nearly twice that for whites, and the gap widened sharply in the 1960's.

*Children per 1000 Adult Males*

|         | 1960  | 1966  |
|---------|-------|-------|
| White   | 1,365 | 1,406 |
| Negro   | 1,922 | 2,216 |

It is this factor, surely, that accounts for much of the present distress of the black urban slums. At the same time, it is fairly clear that the sharp escalation in the number of births that characterized the past twenty-five years has more or less come to an end. The number of Negro females under age five is exactly the number aged five to nine. Thus the 1980's will see a slackening of the present severe demands on the earning power of adult Negroes, and also on the public institutions that provide services for children. But for the decade immediately ahead, those demands will continue to rise—especially for central city blacks, whose median age is a little more than ten years below that for whites— and will clearly have a priority claim on public resources.

*1967 Negro Female Population*

| AGE | NUMBER |
|---|---|
| Under 5 | 1,443,000 |
| 5 to 9 | 1,443,000 |
| 10 to 14 | 1,298,000 |
| 15 to 19 | 1,102,000 |
| 20 to 24 | 840,000 |

**2.** Economic and social forces in urban areas are not self-balancing. Imbalances in industry, transportation, housing, social services and similar elements of urban life frequently tend to become more rather than less pronounced, and this tendency is often abetted by public policies. The concept of urban balance may be tentatively set forth: a social condition in which forces tending to produce imbalance induce counterforces that simultaneously admit change while maintaining equilibrium. It must be the constant object of federal officials whose programs affect urban areas—and there are few whose do not—to seek such equilibrium.

The evidence is considerable that many federal programs have induced sharp imbalances in the "ecology" of urban areas—the highway program, for example, is frequently charged with this, and there is wide agreement that other, specifically city-oriented programs such as urban renewal, have frequently accomplished just the opposite of their nominal objectives. The reasons are increasingly evident. Cities are complex social systems. Interventions that, intentionally or not, affect one component of the system almost invariably affect second, third, and fourth components as well, and these in turn affect the first component, often in ways quite opposite to the direction of the initial intervention.

Most federal urban programs have assumed fairly simple cause and effect relationships which do not exist in the complex real world. Moreover, they have typically been based on "common sense" rather than research in an area where common sense can be notoriously misleading. In the words of Jay W. Forrester, "With a high degree of confidence we can say that the intuitive solution to the problems of complex social systems will be wrong most of the time."

This doubtless is true, but it need not be a traumatizing truth. As Lee Rainwater, Harvard professor of sociology and a contributor to this volume, argues, the logic of multivariate analysis and experience with it suggest that some components of a complex system are always vastly more important than others, so that when (if) these are accurately identified, a process of analysis that begins with the assertion of chaos can in fact end by producing quite concise and purposeful social strategies.

3. At least part of the relative ineffectiveness of the efforts of urban government to respond to urban problems derives from the fragmented and obsolescent structure of urban government itself. The federal government should constantly encourage and provide incentives for the reorganization of local government in response to the reality of metropolitan conditions. The objective of the federal government should be that local government be stronger and more effective, more visible, accessible, and meaningful to local inhabitants. To this end the federal government should discourage the creation of paragovernments designed to deal with special problems by evading or avoiding the jurisdiction of established local authorities, and should encourage effective decentralization.

Although the quality of local government, especially in large cities, has been seen to improve of late, there appears to have been a decline in the vitality of local political systems, and an almost total disappearance of serious effort to reorganize metropolitan areas into new and more rational governmental jurisdictions. Federal efforts to recreate ethnic, neighborhood-based community organizations, as in the poverty program, or to induce metropolitan area planning as in various urban development programs, have had a measure of success, but nothing like that hoped for. The middle-class norm of "participation" has diffused downward and outward, so that federal urban programs now routinely require citizen participation in the planning process and beyond, yet somehow this does not seem to have led to more competent communities. In some instances it appears rather to have escalated the level of stalemate.

It may be we have not been entirely candid with ourselves in this area. Citizen participation, as Elliott A. Krause has pointed out, is in practice a "bureaucratic ideology," a device whereby public officials induce nonpublic individuals to act in a way the officials desire. Although the putative object may be, indeed almost always is, to improve the lot of the citizen, it is not settled that the actual consequences are anything like that. The ways of the officials, of course, are often not those of the elected representatives of the people, and the citizens may become a rope in the tug of war between bureaucrat and representative. Especially in a federal system, citizen participation easily becomes a device whereby the far off federal bureaucracy acquires a weapon with which to battle the elected officials of local government. Whatever the nominal intent, the normal outcome is federal support for those who would diminish the legitimacy of local government. But it is not clear that the federal purposes are typically advanced through this process. To the contrary, an all-round diminishment, rather than enhancement of energies seems to occur.

(This would appear especially true when citizen participation has in effect meant putting citizens on the payroll. However much they may continue to protest, the protest acquires a certain hollow ring. Something like this has surely happened to groups seeking to influence public opinion on matters of public policy when the groups have been openly or covertly supported by the federal government. This is a new practice in American democracy. It began in the field of foreign affairs, and has now spread to the domestic area. To a quite astonishing degree it will be found that those groups which nominally are pressing for social change and development in the poverty field, for example, are in fact subsidized by federal funds. This occurs in protean ways—research grants, training contracts, or whatever—and is done with the best of intentions. But, again, with what results is far from clear. Can this development, for example, account for the curious fact that there seems to be so much protest in the streets of the nation, but so little, as it were, in its legislatures? Is it the case, in other words, that the process of public subsidy is subtly debilitating?)

Whatever the truth of this judgment, it is nevertheless clear that a national urban policy must look first to the vitality of the elected governments of the urban areas, and must seek to increase their capacity for independent, effective, and creative action. This suggests an effort to find some way out of the present fragmentation, and a certain restraint on the creation of federally-financed "competitive governments."

Nathan Glazer has made the useful observation that in London and Tokyo comprehensive metropolitan government is combined with a complex system of "sub-governments"—the London Boroughs—representing units of 200,000–250,000 persons. These are "real" governments, with important powers in areas such as education, welfare, and housing. In England, at all events, they are governed through an electoral system involving the national political parties in essentially their national postures. (Indeed, the boroughs make up the basic units of the parties' urban structure.) It may well be there is need for social inventions of this kind in the great American cities, especially with respect to power over matters such as welfare, education, and housing which are now subject to intense debates concerning local control. The demand for local control is altogether to be welcomed. In some degree it can be seen to arise from the bureaucratic barbarities of the highway programs of the 1950's, for example. But in the largest degree it reflects the processes of democracy catching up with the content of contemporary government. As government more and more involves itself in matters that very much touch on the lives of individual citizens, those individuals seek a greater voice in the programs concerned. In the hands of ideologues or dimwits, this demand can lead to an utter paralysis of government. It has already done so in dozens of urban development situations. But approached with a measure of sensitivity—and patience—it can lead to a considerable revitalization of urban government.

4. A primary object of federal urban policy must be to restore the fiscal vitality of urban government, with the particular object of ensuring that local governments normally have enough resources on hand or available to make local initiative in public affairs a reality.

For all the rise in actual amounts, federal aid to state and local government has increased only from 12 per cent of state-local revenue in 1958 to 17 per cent in 1967. Increasingly, state and local governments that try to meet their responsibilities lurch from one fiscal crisis to another. In such circumstances, the capacity for creative local government becomes least in precisely those jurisdictions where it might most be expected. As much as any other single factor, this condition may be judged to account for the malaise of city government, and especially for the reluctance of the more self-sufficient suburbs to associate themselves with the nearly bankrupt central cities. Surviving from one fiscal deadline to another, the central cities commonly adopt policies which only compound their ultimate difficulties. Yet their options are so few. As James Q. Wilson writes, "The great bulk of any city's budget is, in effect, a fixed charge the mayor is powerless to alter more than trivially." The basic equation, as it were, of American political economy is that for each 1 per cent increase in the Gross National Product (GNP) the income of the federal government increases 1.5 per cent while the normal income of city governments rises .5 to .75 at most. Hence both a clear opportunity and a no less manifest necessity exist for the federal government to adopt as a deliberate policy an increase in its aid to urban governments. This should be done in part through revenue sharing, and in part through an increase in categorical assistance, hopefully in much more consolidated forms than now exist, and through credit assistance.

It may not be expected that this process will occur rapidly. The prospects for an enormous "peace and growth dividend" to follow the cessation of hostilities in Vietnam are far less bright than they were painted. But the fact is that the Gross National Product grows at a better than a billion dollars a week, and we can afford the government we need. This means, among our very first priorities, an increase in the resources available to city governments.

A clear opportunity exists for the federal government to adopt as a deliberate policy an increase in its aid to state and local governments in the aftermath of the Vietnam war. Much analysis is in order, but in approximate terms it may be argued that the present proportion of aid should be about doubled, with the immediate objective that the federal government contribution constitute one-third of state and local revenue.

5. Federal urban policy should seek to equalize the provision of public services as among different jurisdictions in metropolitan areas.

Although the standard depiction of the (black) residents of central cities as grossly deprived with respect to schools and other social services, when compared with their suburban (white) neighbors, requires endless qualification, the essential truth is that life for the well to do is better than life for the poor, and that these populations tend to be separated by artificial government boundaries within metropolitan areas. (The people in between may live on either side of the boundaries, and are typically overlooked altogether.)

As a minimum, federal policy should seek a dollar-for-dollar equivalence in the provision of social services having most to do with economic and social opportunity. This includes, at the top of the list, public education and public safety. (Obviously there will always be some relatively small jurisdictions—e.g., the Scarsdale school system—that spend a great deal more than others, but there can be national or regional norms and no central city should be forced to operate below them.)

Beyond the provision of equal resources lies the troubled and elusive question of equal results. Should equality of educational opportunity extend to equality of educational achievement (as between one group of children and another)? Should equality of police protection extend to equality of criminal victimization? That is to say, should there be not only as many police, but also as few crimes in one area of the city as in another? These are hardly

16

simple questions, but as they are increasingly posed it is increasingly evident that we shall have to try to find answers.

The area of housing is one of special and immediate urgency. In America, housing is not regarded as a public utility (and a scarce one!) as it is in many of the industrial democracies of Europe, but there can hardly be any remaining doubt that the strong and regular production of housing is very nearly a public necessity. We shall not solve the problem of racial isolation without it. Housing must not only be open, *it must be available.* The process of filtration out from dense center city slums can only take place if the housing perimeter, as it were, is sufficiently porous. For too long now the production of housing has been a function not of the need for housing as such, but rather of the need to increase or decrease the money supply, or whatever. Somehow a greater regularity of effective demand must be provided the housing industry, and its level of production must be increased.

**6.** The federal government must assert a specific interest in the movement of people, displaced by technology or driven by poverty, from rural to urban areas, and also in the movement from densely populated central cities to suburban areas.

Much of the present urban crisis derives from the almost total absence of any provision for an orderly movement of persons off the countryside and into the city. The federal government made extraordinary, and extraordinarily successful, efforts to provide for the resettlement of Hungarian refugees in the 1950's and Cuban refugees in the 1960's. But almost nothing has been done for Americans driven from their homes by forces no less imperious.

Rural to urban migration has not stopped, and will not for some time. Increasingly, it is possible to predict where it will occur, and in what time sequence. In 1968, for example, testing of mechanical tobacco harvesting began on the East Coast and the first mechanical grape pickers were used on the West Coast. Hence, it is possible to prepare for it, both by training of those who leave, and providing for them where they arrive. Doubtless the United States will remain a nation of exceptionally mobile persons, but the completely unassisted processes of the past need not continue with respect to the migration of impoverished rural populations. There are increasing indications that the dramatic movement of Negro Americans to central city areas may be slackening, and that a countermovement to surrounding suburban areas may have begun. This process is to be encouraged in every way, especially by the maintenance of a flexible and open housing market.

But it remains the case that in the next thirty years we shall add one hundred million persons to our population. Knowing that, it is impossible to have no policy with respect to where they will be located. To let nature take its course is a policy. To consider what might be best for all concerned and to seek to provide it is surely a more acceptable goal.

7. State government has an indispensable role in the management of urban affairs, and must be supported and encouraged by the federal government in the performance of this role.

This fact, being all but self–evident, tends to be overlooked. The trend of recent legislative measures, almost invariably prompted by executive initiatives, has been to establish a direct federal-city relationship. States have been bypassed, and doubtless some have used this as an excuse to avoid their responsibilities of providing the legal and governmental conditions under which urban problems can be effectively confronted.

It has, of course, been a tradition of social reform in America that city government is bad and that, if anything, state government is worse. This is neither true as a generalization nor useful as a principle. But on the other hand, by and large, state governments, with an occasional exception such as New York, have not involved themselves with urban problems, and are readily enough seen by mayors as the real enemy. But this helps neither. States must become involved. City governments, without exception, are creatures of state governments. City boundaries, jurisdictions, and powers are given and taken away by state governments. It is surely time the federal establishment sought to lend a sense of coherence and a measure of progressivism to this fundamental process.

The role of state government in urban affairs cannot easily be overlooked: it is more typically ignored on political or ideological grounds. By contrast, it is relatively easy to overlook county government, and possibly an even more serious mistake to do so. In a steadily increasing number of metropolitan areas, the county, rather than the original core city, has become the only unit of government that makes any geographical sense. That is to say, the only unit whose boundaries contain most or all of the actual urban settlement. The powers of county government have typically lagged well behind its potential, but it may also be noted that in the few—the very few—instances of urban reorganization to take place since World War II, county government has assumed a principal, even primary role in the new arrangement.

**8.** The federal government must develop and put into practice far more effective incentive systems than now exist whereby state and local governments, and private interests can be led to achieve the goals of federal programs.

The typical federal grant–in–aid program provides its recipients with an immediate reward for promising to work toward some specified goal—raising the educational achievement of minority children, providing medical care for the poor, cleaning up the air, reviving the downtown business district—but almost no reward for actually achieving such goals, and rarely any punishment for failing to do so.

It is by now widely agreed that what federal grant–in–aid programs mostly reward is dissimulation. By and large the approach of the federal government to most urban problems is to provide local institutions with money in the hope they will perform but with no very powerful incentives to do so.

There is a growing consensus that the federal government should provide market competition for public programs, or devise ways to imitate market conditions. In particular, it is increasingly agreed that federal aid should be given directly to the consumers of the programs concerned—individuals included—thus enabling them to choose among competing suppliers of the goods or services that the program is designed to provide.

Probably no single development would more enliven and

19

energize the role of government in urban affairs than a move from the monopoly service strategy of the grant-in-aid programs to a market strategy of providing the most reward to those suppliers that survive competition.

In this precise sense, it is evident that federal programs designed to assist those city-dwelling groups that are least well off, least mobile, and least able to fend for themselves must in many areas move beyond a *services* strategy to an approach that provides inducements to move from a dependent and deficient status to one of independence and sufficiency. Essentially, this is an *income* strategy, based fundamentally on the provision of incentives to increase the earnings and to expand the property base of the poorest groups.

Urban policy should in general be directed toward raising the level of political activity and concentrating it in the electoral process. It is nonetheless possible and useful to be alert for areas of intense but unproductive political conflict and to devise ways to avoid such conflict through market strategies. Thus conflicts over "control" of public education systems have frequently of late taken on the aspect of disputes over control of a monopoly, a sole source of a needed good. Clearly some of the ferocity that ensues can be avoided through free choice arrangements that, in effect, eliminate monopoly control.

If we move in this direction, difficult "minimum standard" regulation problems will almost certainly arise, and must be anticipated. No arrangement meets every need, and a good deal of change is primarily to be justified on grounds that certain systems need change for its own sake. For example, small school districts, controlled by locally elected boards may be just the thing for New York City. However, in Phoenix, Arizona, where they have just that, consolidation and centralization would appear to be the desire of educational reformers. But either way, a measure of market competition can surely improve the provision of public services, much as it has proved an efficient way to obtain various public paraphernalia, from bolt action rifles to lunar landing vehicles.

Here, as elsewhere, it is essential to pursue and to identify the formerly hidden urban policies of government. These are nowhere more central to the issue than in the matter of incentives. Thus for better than half a century now, city governments with the encouragement of state and federal authorities have been seeking to direct urban investment and development in accordance with principles embodied in zoning codes, and not infrequently in accord with precise city plans. However, during this same time the tax laws have provided the utmost incentive to pursue just the opposite objectives of those incorporated in the codes and the plans. It has, for example, been estimated that returns from land speculation based on zoning code changes on average incur half the tax load of returns from investment in physical improvements. Inevitably, energy and capital have diverted away from pursuing the plan, toward subverting it. It avails little for government to deplore the evasion of its purposes in such areas. Government has in fact established two sets of purposes, and provided vastly greater inducements to pursue the implicit rather than the avowed ones. Until public authorities, and the public itself, learn to be much more alert to these situations, and far more open in discussing and managing them, we must expect the present pattern of self-defeating contradictions to continue.

9. The federal government must provide more and better information concerning urban affairs, and should sponsor extensive and sustained research into urban problems.

Much of the social progress of recent years derives from the increasing quality and quantity of government-generated statistics and government-supported research. However, there is general agreement that the time is at hand when a general consolidation is in order, bringing a measure of symmetry to the now widely dispersed (and somewhat uneven) data-collecting and research-supporting activities of the federal government. Such consolidation should not be limited to urban problems, but it must surely include attention to urban questions.

The federal government should, in particular, recognize that most of the issues that appear most critical just now do so in large measure because they are so little understood. This is perhaps especially so with respect to issues of minority group education, but generally applies to all the truly difficult and elusive issues of the moment. More and better inquiry is called for. In particular, the federal government must begin to sponsor longitudinal research designed to follow individual and communal development over long periods of time.

It should also consider providing demographic and economic projections for political subdivisions as a routine service, much as the weather and the economy are forecast. (Thus, Karl Taueber has shown how seemingly unrelated policies of local governments can increase the degree of racial and economic differentiation between political jurisdictions, especially between central cities and suburbs.)

Similarly, the extraordinary inquiry into the education system begun by the U.S. Office of Education under the direction of James S. Coleman should somehow be established on a continuing basis. It is now perfectly clear that little is known about the processes whereby publicly provided resources affect educational outcomes. The great mass of those involved in education, and of that portion of the public which interests itself in educational matters, continue undisturbed in the old beliefs. But the bases of their beliefs are already thoroughly undermined and the whole structure is likely to collapse in a panic of disillusion and despair unless something like new knowledge is developed to replace the old. Here again, longitudinal inquiries are essential. And here also, it should be insisted that however little the new understandings may have diffused beyond the academic research centers in which they originated, the American public is accustomed to the idea that understandings do change and, especially in the field of education, is quite open to experimentation and innovation.

Much of the methodology of social science originated in clinical psychology, and perhaps for that reason tends to be deficiency-oriented. Social scientists raise social problems, the study of which can become a social problem in its own right if it is never balanced by the identification and analysis of social successes. We are not an unsuccessful country. To the contrary, few societies work as hard at their problems, solve as many, and in the process stumble on more unexpected and fulsome opportunities. The cry of the decent householder who asks why the profession (and the news media which increasingly follow the profession) must be ever preoccupied with juvenile delinquency and never with "juvenile decency" deserves to be heard. Social science, like medical science, has been preoccupied with pathology, with pain. A measure of inquiry into the sources of health and pleasure is overdue, and is properly a subject of federal support.

10. The federal government, by its own example, and by incentives, should seek the development of a far heightened sense of the finite resources of the natural environment, and the fundamental importance of aesthetics in successful urban growth.

The process of "uglification" may first have developed in Europe, but as with much else, the technological breakthroughs have taken place in the United States. American cities have grown to be as ugly as they are, not as a consequence of the failure of design so much as of the success of a certain interaction of economic, technological, and cultural forces. It is economically efficient to exploit the natural resources of land, air, and water by technological means which the culture does not reject, even though the result is an increasingly despoiled, debilitated, and now even dangerous urban environment.

It is not clear how this is to change, and so the matter which the twenty-second century, say, will almost certainly see as having been the primary urban issue of the twentieth century, is ranked last in the public priorities of the moment. But there are signs that the culture is changing, that the frontier sense of a natural environment of unlimited resources, all but impervious to human harm, is being replaced by an acute awareness that serious, possibly irreparable harm is being done to the environment, and that somehow the process must be reversed. This could lead to a new, nonexploitive technology, and thence to a new structure of economic incentives.

The federal establishment is showing signs that this cultural change is affecting its actions, and so do state and city governments. But the process needs to be raised to the level of a conscious pursuit of policy. The quality of the urban environment, a measure deriving from a humane and understanding use of the natural resources, together with the creative use of design in architecture and in the distribution of activities and people must become a proclaimed concern of government. And here the federal government can lead. It must seek out its hidden policies. The design of public housing projects, for example, surely has had the consequence of manipulating the lives of those who inhabit them. By and large the federal government set the conditions which have determined the disastrous designs of the past two decades. It is thus responsible for the results, and should force itself to realize that. And it must be acutely aware of the force of its own example. If scientists (as we are told) in the Manhattan Project were prepared to dismiss the problem of long-lived radioactive wastes as one that could be solved merely by ocean dumping, there are few grounds for amazement that business executives in Detroit for so long manufactured automobiles that emitted poison gases into the atmosphere. Both patterns of decision evolved from the primacy of economic concerns in the context of the exploitation of the natural environment in ways the culture did not forbid. There are, however, increasing signs that we are beginning to change in this respect. We may before long evolve into a society in which the understanding of and concern about environmental pollution, and the general uglification of American life, will be both culturally vibrant and politically potent.

Social peace is the primary objective of social policy. To the extent that this derives from a shared sense of the value and significance of the public places and aesthetic value of the city, the federal government has a direct interest in encouraging such qualities.

Daniel J. Elazar has observed that while Americans have been willing to become urbanized, they have adamantly resisted becoming citified. Yet a measure of this process is needed. There are not half a dozen cities in America whose disappearance would, apart from the inconvenience, cause any real regret. But to lose one of those half-dozen would plunge much of the nation and almost all the immediate inhabitants into genuine grief. Something of value in our lives would have been lost, and we would know it. The difference between those cities that would be missed and those that would not be resides fundamentally in the combination of architectural beauty, social amenity, and cultural vigor that so sets them apart. It has ever been such. To create such a city and to preserve it was the great ideal of the Greek civilization, and it may yet become ours as we step back ever so cautiously from the worship of the nation state with its barbarous modernity and impotent might. We might well consider the claims for a different life asserted in the oath of the Athenian city-state:

We will ever strive for the ideals and sacred things of the city, both alone and with many;
We will unceasingly seek to quicken the sense of public duty;
We will revere and obey the city's laws;
We will transmit this city not only not less, but greater, better and more beautiful than it was transmitted to us.

# 2 POPULATION COMPOSITION AND TRENDS

## Philip M. Hauser

Among the major developments which have shaped man's attitudes, values, and behavior are the population explosion, the population implosion, and population diversification. The population explosion is the remarkable acceleration in rates of population growth, especially during the three centuries of the modern era. The population implosion is the increasing concentration of population in urban and metropolitan areas—in ever smaller proportions of the earth's land surface. The population implosion has, of course, been fed by the population explosion; and both the population explosion and implosion have generated population diversification—the increasing heterogeneity of peoples, by culture, values, religion, ethnicity, and race, who share the same geographic area and, increasingly, the same life space.

Most of the problems afflicting the contemporary world, in the economically advanced or in the developing areas, can be better comprehended when viewed as frictions arising from these relatively recent phenomena. In both the economically advanced and the developing notions, despite great differences in the extent of urbanization, major physical, personal, social, economic, and governmental problems are evidences of the transition, still under way, from an agrarian to an urban and metropolitan order. Man, as the only culture-building animal on the face of the earth, not only adapts to environment, he creates an environment to which to adapt. And he is still learning to live in the urban and metropolitan world he is creating.

The United States is one of history's most dramatic examples of all three developments—the population explosion, the population implosion, and population diversification—and many of the contemporary problems of this nation, as in the rest of the world, can be traced to them.

## TOTAL POPULATION GROWTH

In 1790, when the first decennial census of the United States was taken, the United States had a total population of less than 4 million persons. Between 1790 and 1950 the population of the nation doubled five times to reach a total of 151 million. The first three times, between 1790 and 1865, each doubling took twenty-five years. To double the fourth time required thirty-five years, from 1865 to 1900; and for the fifth doubling it took the first half of this century. With the help of the postwar baby boom it is possible that a sixth doubling, giving a population of over 300 million persons, will occur before the end of this century. The increased time required for doubling up to 1950 indicates, of course, the slackening pace of population growth initiated during the latter part of the nineteenth century and continuing until World War II. Under the impact of the postwar conditions the rate of the population growth, as of many other nations, accelerated. By 1960 the population of the nation numbered more than 180 million; during 1967 it reached 200 million.

The rapid population growth of the United States was the product both of natural increase, the excess of births over deaths, and of immigration. Between 1820, when the government first began to count newcomers, and 1966, almost 44 million immigrants were admitted into the country, predominantly from Europe. The peak in immigration was reached during the decade 1901 to 1910, when some 8.8 million immigrants were admitted. During World War I immigration slackened, and then, with the passage, beginning in the 1920's, of the various laws controlling the admission of newcomers, immigration became a minor factor in population growth. It is now restricted to about 300,000 immigrants per year. Despite the relatively large volume of immigration, in no decade did immigration exceed natural increase in contributing to total population growth.

Natural increase, including of course that of the immigrants, has been the main source of population growth of the nation. The birth rate of the United States in 1800 has been estimated to have been as high as 55 (births per 1,000 persons per year), a level that matches the highest birth rates ever achieved by any nation. Birth rates in the developing nations today are, generally, at levels above 40 and some above 50. Death rates in the United States in 1800 were also high, however, probably approximating a level close to 30 (deaths per 1,000 persons per year). A birth rate of 55 together with a death rate of 30 could produce a rate of natural increase of 25, or a population growth rate, by reason of natural increase alone, of 2.5 per cent per year. This is a level of growth not too different from that of India today.

Since 1800, however, both the birth rate and the death rate of the United States have declined considerably. The birth rate fell, with some cyclical and other fluctuations, from the rate of 55 in 1800 to a low of about 18 during the Depression 1930's. It began to rise as economic recovery was experienced in the late 1930's, and experienced a sharp upturn after World War II to average a level of about 25 from 1947 to 1958. It has been declining ever since, but the decline is more the result of the changing age structure than of any decrease in number of children being born per couple. The annual birth rate, based on the relationship between total number of births and total population does not necessarily reveal whether there has been any change in "cohort fertility," that is the number of children born per family during the entire reproductive period. At the present time it is not yet known whether, after a long period of decline, the increase in the number of children born per couple, which began with women born around 1910, has run its course and has again begun to decline.

In any case, the absolute number of births, which exceeded 4 million per year from 1954 through 1964 and fell below that level since, is almost certain to rise above 4 million again before the end of this decade and is likely to exceed 5 million per year in the early 1970's. The reason for this probability is to be found in the fact that the number of persons of reproductive age in the United

States will about double in the coming generation as the postwar babies reach marriageable age.

Moreover, the death rate has also dropped sharply since 1800, although most of the gain in longevity has occurred during this century. From a level of perhaps 17 in 1900, the death rate declined to a level below 10 by 1948 and, it is noteworthy, has changed very little since.

By reason of the changes in fertility and mortality the rate of natural increase in the United States declined from about 25 (excess of births over deaths per 1,000 persons per year) in 1800 to a level of about 7 at the bottom of the Depression. With the postwar baby boom, natural increase rose to a level of about 15, but it declined again to a level of about 9 in 1966. Thus, the excess of births over deaths alone, without considering immigration, has changed over time so as to drop from a contribution of a 2.5 per cent annual growth rate in 1800, to a .7 per cent growth rate in 1935, to a 1.5 per cent growth rate during the postwar baby boom, to a .9 per cent growth rate in 1966. Although changes in the birth and death rates considered (the "crude" rates) reflect in considerable part changes in the nation's age structure and do not accurately depict what has happened to the level of child bearing per couple during the entire reproductive span, a more complex subject not considered here, they do show how the net effect of fertility and mortality changes have operated to contribute to the decline in the rate of total population growth of the United States.

The U.S. Bureau of the Census has from time to time made projections of U.S. population on varying assumptions about the future course of fertility and mortality. Such projections made in 1967 indicate that, despite the declining crude birth rate, the United States will continue to experience large absolute population increases in the decades which lie ahead. These projections show that by 1990, only 20 years hence, the population of the United States may reach a level of from 256 to 300 million. One of these projections, based on the assumption that fertility would remain at the level obtaining in 1964 and 1965, would produce a population of 207 million by 1970, 243 million by 1980, and 287 million by 1990. The same projection gives a population of 336 million by the year 2000 and 430 million by 2015.

## URBANIZATION

In 1790, 95 per cent of the population of the United States lived in rural areas, that is, on farms or in places having fewer than 2,500 persons. The 5 per cent of the population who lived in cities were concentrated in 24 such places, only two of which (New York and Philadelphia) had populations of 25,000 or more. By 1850, population in urban places was still as low as 15 per cent. By 1900, however, almost two-fifths of the population lived in cities. But it was not until as recently as 1920 that the United States became an urban nation in the sense that more than half of the population lived in cities. That many critical problems affect cities and urban populations should not be too surprising in light of the fact that it will not be until the next census of population is taken in 1970 that the United States will have completed her first half-century as an urban nation.

The increase in urban and metropolitan population is the result of net migration as well as natural increase. Cities and metropolitan areas have over the years received large numbers of migrants from rural and nonmetropolitan areas of the United States as well as through immigration from abroad. For example, between 1950 and 1960, 35 per cent of the total metropolitan growth was the result of net migration (including immigration) and 65 per cent the result of natural increase.

Migration, in the United States, as elsewhere, represents mainly a movement of population from places of lesser economic opportunity to places of greater opportunity. Moreover, in the United States, as elsewhere, migrants have often been ill-prepared in their areas of origin for life in these areas of destination. In consequence, the problems of adjustment of in-migrants to urban and metropolitan areas are often difficult as they seek to accommodate to their new setting. Furthermore, the problems of adjustment are compounded when complicated by differences of language, culture, religion, ethnicity, or race.

The speed of the population concentration in urban and metropolitan areas becomes clear in an examination of developments since the turn of the century. In the first sixty years of this century the increase in urban population absorbed 92 per cent of the total population growth in the nation. In the decade 1950 to 1960 the increase in urban population absorbed more than 100 per cent of total national growth; that is, total rural population, including nonfarm as well as farm, actually diminished for the first time.

The population increase in large metropolitan areas is equally dramatic. The increase in the population of the Standard Metropolitan Statistical Areas (SMSA's), as they are officially designated by the federal government (cities of 50,000 or more together with the counties in which they are located), absorbed 85 per cent of total national growth between 1900 and 1960; and the 24 largest SMSA's, those with one million or more, absorbed 48 per cent, almost half of the total growth of the nation in the first sixty years of this century.

In consequence, by 1960, 70 per cent of the American people, 125 million, resided in over 6,000 urban places; and 63 per cent, or 113 million persons, lived in 212 SMSA's. In 1965, it is estimated by the U.S. Bureau of the Census, 65 per cent of the population, or 126 million persons, resided in 222 SMSA's.

The trend towards increased urban and metropolitan concentration of population is likely to continue. The reasons for this are to be found in the advantages of clumpings of population and economic activities. As Adam Smith noted in *The Wealth of Nations,* the greater the agglomeration the greater is the division of labor possible; and this generates increased specialization, easier application of technology and the use of nonhuman energy, economies of scale, external economies, and minimization of the frictions of space and communication. In brief, the trend towards urbanization and metropolitanization is likely to continue because such clumpings of people and economic activities constitute the most efficient producer and consumer units yet devised.

While the population of the United States has become increasingly concentrated in urban and metropolitan areas, population decentralization has occurred within metropolitan areas. That is, within metropolitan areas the proportion of residents living in the suburban ring, the area outside the central city but within the

SMSA has increased. During the first sixty years of this century the increase in the population of central cities absorbed 25 per cent of total national growth and between 1950 and 1960, 31 per cent. In contrast the increase in suburban population absorbed 45 per cent of total national growth between 1900 and 1960, and 66 per cent, a full two-thirds, of total national growth in the decade 1950 to 1960.

As a result, by 1960 almost half of the total population in SMSA's lived in the suburban ring; the central cities contained but a slight majority of metropolitan population. By 1965, however, it is estimated by the Census Bureau that suburban population exceeded that of the central cities. In 1965, suburban ring population is estimated at 65 million, or 51.9 per cent of the metropolitan population, while central city population is placed at 60 million, or 48.1 per cent.

The reason for the decentralization of population within metropolitan areas is not difficult to trace. It is not so much the result of "flight from the city" as the joint effect of the following two factors. First, with advancing technology the maximum possible size of a metropolitan area has continuously increased. Twentieth-century technology, characterized by electric power, the combustion engine complex—the auto, truck, and highway—and the telephone, has made possible much larger clumpings of people and economic activities than any prior technologies. Second, the central city in the United States is a creature of the state legislature which incorporates it, grants its charter, delimits its powers, and defines its boundaries. Although cities have some powers of annexation, the rate of population growth has far exceeded the rate of annexations. The reason, then, why suburban rings are growing faster than ceneral cities is simply that with the increased size of metropolitan areas and the historical fact that most central cities have been filled up since 1920, the only place that additional growth could occur was in suburbia.

Short of catastrophic events such as nuclear war, it may be predicted that urban and metropolitan concentration will continue. Within the framework of total population growth outlined above and the assumption of the continuation of urban and metropolitan trends, it is possible that by 1990 the United States, with a total population of 287 million, would have 233 million urban residents, 81 per cent of the total, and 199 million metropolitan residents, about 70 per cent of the total. Suburban rings could contain some 119 million persons, or about 60 per cent of the metropolitan population; while central cities could hold some 80 million persons, or 40 per cent of the total. Perhaps 116 million persons, or 58 per cent of the total national population, will be resident in the large SMSA's having a million or more persons.

## ETHNIC AND RACIAL COMPOSITION

The United States has been one of history's most dramatic examples of population diversification as well as of the population explosion and the population implosion. Although the original European settlers were predominantly from the United Kingdom, the infusion of African Negro population began during the eighteenth century and was followed by waves of diverse European stocks during the nineteenth and early twentieth centuries.

The census of population first counted "foreign born" whites in 1850. At that time they constituted 9.7 per cent of the total population. Although successive waves of immigration were heavy, the foreign-born whites never exceeded 14.5 per cent of the total, a level reached in 1890 and again in 1910. They have been a dwindling proportion of the total ever since 1910. By reason of restrictions on immigration, the foreign-born will become a decreasing proportion of the population of the nation in the decades which lie ahead.

As has been indicated, between 1820 and 1966, some 44 million immigrants, mainly from Europe, entered the United States. The predominant proportion of immigrants settled in the nation's cities. The immigrants came in great waves. During the nineteenth century large streams of Irish, Germans, and Scandinavians were admitted following crop failures, hard economic times, or political difficulties in their countries of origin. During the early part of the twentieth century the sources of immigration shifted from Western and Northern, to Eastern and Southern, Europe. Large numbers of Russians and Poles, including Jews, Italians, Bohemians, Greeks,

33

and other peoples responded to the opportunities open to them in the United States.

The processes by which these successive waves of newcomers made their entry into the metropolitan United States, found residential locations, acquired jobs, and achieved status in the social order were strikingly uniform.

Because the cheapest dwelling units in American cities were located in their inner zones, the newly arrived immigrants found their ports of entry and areas of first settlement there, in the older and blighted areas of the city. They initially worked at the least desirable, menial, and lowest paid occupations. They each in turn had the lowest social status and were greeted with suspicion, distrust, prejudice, and discriminatory practices on the part of those who had come earlier. For example, each of the immigrant newcomer groups was greeted with derisive designations. During the nineteenth century the newcomers were known as "Micks" (the Irish), "Krautheads" (the Germans), or "dumb Swedes" (the Scandinavians). During the twentieth century they were known as "Polaks" (the Poles), "Sheenies" (the Jews), "Wops" (the Italians), "Bohunks" (the Bohemians), and so on.

With the passage of time, each wave of the white ethnic immigrants climbed the social and economic ladder as measured by place of residence, job and remuneration, and social acceptability. Each of the white ethnic groups had the option of continuing to live in neighborhoods of their own or in dispersed and integrated fashion. In general, the longer their residence in the United States, the larger is the proportion who have elected to leave their enclaves and live in integrated fashion.

Needless to say, the assimilation of these immigrant groups, popularly known as "Americanization," was not achieved without a number of frictions. Although the United States is often referred to as a "melting pot," it is clear that assimilation of the immigrants has not proceeded either smoothly or uniformly—in fact the process is still very much under way, and accounts for some of the acute problems of intergroup relations which the nation still experiences.

The more recent newcomers to the urban and metropolitan areas are "in-migrants" rather than "immigrants." That is, they come from the rural and less developed parts of the United States itself, rather than from abroad. The visible Negro and the less visible rural white, including the Appalachian white or "hillbilly," have replaced white immigrants as the main source of new urban and metropolitan settlers. These groups are supplemented by smaller streams of Mexicans, Puerto Ricans, and Orientals as well as the greatly reduced numbers of white immigrants from Europe.

Because of their special problems, it is useful to examine the population trends of Negro Americans. In 1790, as recorded in the first census of the United States, there were less than 800,000 Negroes in the nation, but they made up about 20 per cent of the total population. By that date they had already been resident in the colonies for 175 years, mainly as the property or indentured servants of their white masters.

Negro Americans remained about one-fifth of the total population until 1810. From then to 1930 they were an ever declining proportion of the total, as slave traffic ceased and white immigration continued. By 1930 the proportion of Negroes had diminished to less than one-tenth of the total. Since 1940, however, the Negro growth rate has been greater than that of the white population, and their proportion had risen to 11 per cent by 1967.

In 1790, 91 per cent of all Negroes lived in the South. The first large migratory flow of Negroes out of the South began during World War I, prompted by the need for wartime labor and the freeing of the Negro from the soil, with the diversification of agriculture and the onset of the delayed industrial revolution in the South. This migration of Negroes from the South was greatly increased during and after World War II. As a result the proportion of total Negroes located in the North and West almost quadrupled between 1910 and 1960, increasing from 11 to 40 per cent.

The migratory movement of Negroes from the South to the North and West effected not only a regional redistribution but also, and significantly, an urban-rural redistribution. In 1910, before the out-migration of the Negro from the South began, 73 per cent lived in rural areas—on farms or in places having fewer than 2,500 persons. By 1960, within fifty years, less than a lifetime, the Negro had been transformed from 73 per cent rural to 73

35

per cent urban, and had become more urbanized than the white population.

The great urban concentration of Negro Americans is even more dramatically revealed by their location in metropolitan areas. In 1910, only 29 per cent of Negroes lived in the Standard Metropolitan Statistical Areas. By 1960, this concentration had increased to 65 per cent. By 1960, 51 per cent of all Negroes lived in the central cities of the SMSA's. Moreover, the 24 SMSA's with one million or more inhabitants contained 38 per cent, and their central cities 31 per cent, of all Negro Americans.

By reason of the above developments, by 1850, native whites made up 74.6 per cent of the population of the nation, foreign-born whites 9.7 per cent, and "nonwhites," mainly Negroes, 15.7 per cent. By 1900, the proportions had changed little: 75.5 per cent being native white, 13.4 per cent foreign-born, and 12.1 per cent nonwhite. But in 1900, little more than half the American people were native whites of native parentage. That is, about one-fifth of the population was "second generation," or native whites born of foreign or mixed parentage.

By 1960, native whites constituted 83 per cent, foreign whites 5.2 per cent, and Negroes 10.6 per cent of the total. Native whites of native parentage made up 70 per cent of the total, the remaining 13 per cent of native whites being second generation. Thus, in 1960, "foreign white stock," (foreign-born plus second generation) still made up over 18 per cent of the total population.

Although the foreign white stock will become a dwindling part of the population in the decades which lie ahead, the proportion of nonwhites, mainly Negroes, is likely to increase. In 1960, there were 20.7 million nonwhites in the U.S., or 11.4 per cent of the total. By 1990 it is estimated by the U.S. Bureau of the Census that nonwhites will double, increasing to 41.5 million. By 1990, nonwhites may, therefore, make up some 14.5 per cent of the American people.

The difficult problems of white-Negro relationships have been worsened by the population changes described. The large increase in the population of Negro Americans in urban and metropolitan areas over a relatively short period of time, and the contrasts in background and life styles between Negroes and whites by reason of the disadvantaged position of Negro Americans over the years, have combined to generate tensions that may well constitute the most serious domestic problem of the United States for some time to come.

## CONCLUDING OBSERVATIONS

There have, of course, over time been many other population changes in the United States. The age structure of the American people has profoundly changed, as a result mainly of decreasing fertility, but also of declining mortality. In 1790 the average American was under 16 years old. By 1940 he was over 30; and by 1960 he was, because of the postwar baby boom, again under 30. The proportion of "senior citizens," those 65 years and older, has greatly increased, rising from 3.2 per cent in 1870 to over 9 per cent in 1960. The number of households has continued to grow more rapidly than total population, as the average size of households has decreased—in response to the transition from the extended family system or multinuclear family to the nuclear family, as well as to decreased fertility. Labor force participation of young persons under 20 and of men 65 and over has greatly declined, while that of women has increased in almost like amount. School enrollment has increased, as has years of school completed. Age at marriage has declined and the proportion of the population ever married has increased. There are many more changes which could be greatly elaborated.

Also to be observed are the great differences which have emerged between the urban and the rural population. The urban population, as compared with the rural, tends to be more female and to have a higher median age but smaller proportions of older persons, lower birth rates, a smaller percentage of native whites, lower marriage rates, smaller family size, more formal education, a larger proportion of workers in white-collar occupations, and higher incomes.

To use the language of the late Professor Louis Wirth, urbanization and metropolitanization have produced a new way of life, "urbanism as a way of life," which is still in process of develop-

37

ment. In the transition from an agrarian to an urban society, many problems have been generated, most of which still remain to be resolved. But, along with difficult problems, it must be emphasized that the urban way of life has opened new vistas of opportunity to mankind, higher levels of living, new horizons of the arts and sciences, and ever greater promise of more complete control both of nature and of man's destiny.

# 3 INTERGOVERNMENTAL RELATIONSHIPS IN AN URBANIZING AMERICA

Robert Wood

Urbanization is a force that touches all the countries of the world. To be sure it reaches them in differing dimensions and varying ways, but it is rushing across the globe with a strength that engulfs nations large and small, old and new. Growing numbers of people —too many of them destitute—are relentlessly moving to urban centers for opportunities and for the promise of a better life. In many countries, national development is urban development and the future of growing cities, towns, villages, and hamlets will determine the future of the nation.

## URBANIZATION: THE CHALLENGE OF A TRANSIENT ENVIRONMENT

In the past, man's environment changed almost imperceptibly over the lifetime of many men. Generations of craftsmen worked to complete the great European cathedrals. It took centuries for new ideas to spread across the world. Today, a single man in his lifetime confronts a constantly changing environment.

Not only is this change rapid but its rate is also steadily increasing. Most of us began life before the advent of computers, transistors, satellite communications, or transcontinental airplanes. Textbooks just six years old omit 39 countries. Between 1900 and 1950 more discoveries and inventions were made in the natural sciences and engineering than in all of previously recorded history. Between 1950 and 1960 a similar number of innovations occurred.

## Urban Dimensions in America

It is in this new context of accelerating change and rapid growth that Americans are trying to adapt themselves, to influence their surroundings. Today, the United States is on the verge of an indefinite period of population increase. It has passed the 200 million mark, moving from 76 million at the turn of the century. Within fifty years, 320 million of our 400 million people will live in cities. This geometric increase, this sheer size, this shift in the scale of our continental democracy—more than anything else—is the most significant fact of our urban future.

Our urban residents will need as many more houses, schools, apartments, commercial buildings, parks, recreational areas, and public facilities as have been built since America was colonized 350 years ago. By 1975, twenty-two and one-half million new homes must be built. In that year the United States will need schools sufficient to provide quality education for 60 million children, health and welfare programs for 28 million people over the age of 60, and transportation facilities sufficient to provide daily movement of 200 million people in 90 million cars.

America's expanding population is shifting rapidly and unevenly, compounding the task of orderly development and imposing different requirements on various parts of our country. Thus, national projections understate the dimensions of the problem because growth is not only swift, it is also irregular. With the introduction of mechanized farming the rural poor—often of minority races—began to migrate to the blighted reception areas of our central cities. Dense neighborhoods become centers of deprivation and decay, breeding alienation and apathy which erupt into violence. The small country towns these migrants left behind stagnate and decline.

Moreover, our mobile people tend to cluster in great metropolitan goliaths. In mid-1966, the Baltimore metropolitan area contained 55 per cent of Maryland's population, Chicago, 62 per cent of Illinois', Boston, 59 per cent of Massachusetts', Detroit, 48 per cent of Michigan's.

## THE POLITICS OF MULTI-GOVERNMENT PROBLEM SOLVING

To resolve these complex dilemmas, the American nation like others in the world must establish priorities, mobilize resources, begin planning, build institutions, and carry out programs. In the American political system the development of solutions to urban problems is "institutionally complex"—that is, it involves the efforts of different bodies, public and private, financial and educational—unique to a federal system of government operating with a free economy. Priorities and allocations are, therefore, decisions influenced by the three levels of government established in American federalism and by the many interests of the private sector. The legislative and administrative compromise of these conflicting concerns is the basic manner of problem solving in our country.

### The Multi–Government Setting

The workhorses of American politics are state and local governments. These lower echelons of the American federal system carry out the greatest volume of public business, settle the largest number of political conflicts, make the majority of public decisions, and direct the bulk of public programs. By law and tradition they have major responsibility for maintaining domestic order, for educating our children, for caring for the chronically sick and invalided. With the federal government, they share in the regulation of private business and commerce; they oversee the use and disposition of all property. Their courts resolve most civil and criminal disputes.

The problems of an urbanizing America are being faced today by 81,303 different formal public bodies with specific geographic jurisdictions and substantial independent power: states, counties, cities, towns, boroughs, villages, special districts, and authorities.

These are by no means petty bodies. State and local governments are expanding their activities faster than the federal government. Between 1950 and 1962 their general expenditures increased by 160 per cent, while federal expenditures rose 135 per cent. In 1964 all state and local agencies together had two and one-half times as many employees as the federal government's domestic agencies. By 1975 state and local government employment will increase by 69 per cent, while federal employment will grow by 8 per cent.

States and localities are bound to the federal government by the national Constitution, a similar body of law, national political parties, financial grants and loans, and common constituents who deal simultaneously with all entities in the network. Under the Constitution, the states are the key units, endowed with all governmental powers not vested specifically in the federal government or reserved to the people. All other jurisdictions, whose powers depend on their charters or statute, are creatures of the states.

## The Intergovernment Operation

It is not easy to define the division of functions in the American political system. Political action takes place among and between all parts of the whole. Separate systems and levels are interdependently related.

Morton Grodzins, the political scientist, has pointed out:

> The multitude of governments does not mask any simplicity of activities. There is no neat division of functions among them. If one looks closely, it appears that virtually all governments are involved in virtually all functions. More precisely, there is hardly any activity that does not involve the federal, state, and some local governments in important responsibilities. Functions of the American governments are shared.

President Johnson made this point very clear in his 1967 State of the Union message. Nothing can be achieved by the federal government to settle urban issues, he said, without the cooperation of the states, the cities, and the counties. This theme of interdependency, which touches all problems of American government, is especially important in the urban perspective, where public problems are so obvious and so acute.

## Vertical and Horizontal Axes of Interdependency

Political scientists speak of three types of intergovernmental politics arising from the "interdependent" principle framed in the national Constitution.

First is the relationship among governments born of legal authority and political practice. This is, for instance, the action of each state in carrying out constitutional or statutory changes like amending the Constitution or supervising elections.

Second are the vertical or specialized relationships among bureaucracies of the various government levels. This would, for instance, link city, county, state, and federal administrations concerned with highway construction and road maintenance.

Third—and growing most rapidly—are the horizontal relationships among public bodies at the same level, or in other words, the network of parapolitical local intergovernmental arrangements. In this category we would find efforts like the Ohio Valley Sanitation Compact, the Port of New York Authority, metropolitan-wide councils of government, the U.S. Governor's Conference.

## The Vertical System: Administration by Persuasion

While it is sometimes argued that the vertical axis of federal, state, and local bureaucracies operates as a unitary administration bound by the same programs and policies, administrators function in a more impressionistic landscape. The legal structure, different programs, and distinct professional skills and views stimulate competition that divides and constrains power. The political geography is one of many separate centers often in continuing conflict but related to each other because of the very limits on the participants' power. In this environment, hierarchical and authoritarian techniques of administration are ineffective. The key tool of the American bureaucrat intent on solving a particular problem through intergovernmental channels is persuasion.

## Expanding Systems

Under the stress of urbanization, states and local governments are greatly extending their specialist capacities to deal with resulting social and economic problems. More than twenty states have inaugurated aggressive, resource-oriented urban affairs and community development programs aimed at providing urban services. At least four have undertaken their own housing programs. Others have established offices for continuing attention, review, and assistance on local, legal, financial, organizational, and planning difficulties. Eight states have created cabinet-level departments of urban or community affairs.

In the states and cities there is increasing recognition of the need to reform overlapping jurisdictions and eliminate archaic administrative practices. A number of states have revised their constitutions. Several cities are making innovative organizational changes. Moreover, judicial decisions requiring states to insure equal representation for urban areas in state legislatures have made the urban voter more powerful. One can expect that "vertical" activity will grow because of the incentives and support to state action from federal programs, and as a result of the mounting demand of localities for greater state action.

## The Horizontal System: Centralizing Dependencies

The horizontal system of intergovernmental relations reflects the coming together of different local groups and private associations to deal cooperatively with new and often continuing problems. Usually these joint consultative arrangements are informal and not established by law, though they may also be fixed by statute. Characteristic of this approach are the special purpose authorities or large special districts and public and quasi-public entities active in planning urban development. This tendency towards metropolitan federalism recognizes the actual dependence of all of the participants.

## INTERGOVERNMENTAL RELATIONS IN A
## CHANGING AMERICA

The rapid growth of vertical and horizontal arrangements since the end of World War II represents the pragmatic response of the federal system to the special pressures and problems of urbanization. Thus, a description of the formal structure of the American government is less useful in understanding actual allocation of responsibilities than is an outline of these informal relations.

Governmental organization is no longer a pyramid of building blocks arranged with due respect to a neat and tidy hierarchy. It is viewed and managed as a network of many parts and many relationships. It is viewed not so much in terms of legal prescriptions, authority, rules and regulations, as in terms of the capability to define and solve problems. This has occurred partly because the universe in which intergovernmental arrangements are developed has altered and a new kind of political context is evolving.

### The Politics of Innovation

The latter part of the nineteenth century and the first decades of the twentieth saw Americans struggle to divide scarce resources and benefits among competing classes and interests. Most middle-aged Americans raised in the decade of the Depression and New Deal are accustomed to what may be called the politics of distribution: the concern with satisfying needs for goods and services, with allocating a fixed amount of resources between conflicting groups such as business and labor or farm and city.

Distributive politics was content with making decisions and enforcing them within a stable field. Characteristic of distributive politics is ideology—a doctrine promising to join change with cohesion, progress with purpose, abundance with social justice. These are illusionary ways of dealing with a rapidly changing environment. Since the 1950's, America has turned away from distributive politics. It has concentrated instead on generating fresh resources and moving toward the solutions of problems

45

through the application of knowledge, technology, and new capabilities. The change came slowly, first in the field of defense, then in the application of nuclear energy, and today in the areas of education, welfare, services, and city-building.

The emphasis now is on the politics of innovation: the effort to adjust or create agencies and procedures in such a way that they are responsive and adaptable and that they use and build upon the "uncommon knowledge" made available by modern science which touches the future as well as the past. It no longer suffices simply to absorb and act according to the "common knowledge" of the present. History, experience, custom, and tradition have become poor guides to comprehending our present difficulties—or undertaking to solve them.

Former Vice-President Hubert H. Humphrey often told the story of how during America's hard times, in his home state of South Dakota an entire town would turn out to watch a federal tree planting project aimed at stopping the ravaging dust storms which roared over the prairie. "We all remember what a great day it was for us, we knew at last somebody cared. The trees soon died, but not the memory."

Innovative politics seeks a new combination of resources that will yield new solutions to previously intractable problems. It would care enough to find the method for stopping the dust storms. As Max Ways has written:

> The trouble with many such New Deal "experiments" was that they weren't. They provided material for gladiatorial contests between enthusiastic "have-nots" and "haves." The excitement generated by these political circuses buttered no parsnips and represented a disgraceful waste of the *intellectual* resources of a society that, even then, had scientific and administrative experts who could have been organized for more effective planning. There are, of course, social and moral values (as well as a political value) in letting disadvantaged people know that their government "cares." But there are higher social and moral values in caring enough to make programs work.

## CREATIVE FEDERALISM: THE POLITICS OF
## MULTI-INSTITUTION PROBLEM SOLVING

The continuing evolution of the American system generated by this basic change in political approach is accelerated by the type of problems created by urbanization. The vast orders of magnitude, the size and scope, the complexities of our difficulties dictate solutions that enfold but also exceed intergovernmental relationships; that require specific or local rather than universal or national responses; and that embrace the private as well as the public sectors of our society. Because of this situation, the federal system is going beyond a single approach to problem solving and is developing a multi-institutional method. With a premium on innovation America is moving toward a new period of reconstructing its political institutions. Creative federalism has become the term for describing this phenomenon.

### New—Not Scarce—Power

The growing interplay in the vertical and horizontal relations of public bodies stimulated by urban development has contributed new strength and resources to state and local governments. But, in the terms of traditional allocations of authority and responsibilities, this has not meant a diminution of federal power. Today, federalism does not mean a finite quantity of power being distributed among various governing bodies. The old dual federalism whereby the transfer of authority and resources from one legal body to another was deemed to decrease correspondingly the authority and resources of fellow participants is anachronistic.

In the long American dialogue over states' rights it has been tacitly assumed that the total amount of power was constant, and, therefore, any increase in federal power diminished the power of the states and/or "the people." Creative federalism starts from the contrary belief that total power—private and public, individual and organizational—is expanding very rapidly. As the range of conscious choice widens, it is possible to think of vast increases of federal govern-

47

ment power that do not encroach upon or diminish any other power.

In short, what is happening is not the allocation of old power but the effective distribution of new power based on new energies being released in all sectors of activity.

The emphasis on new energies in all sectors distinguishes creative federalism from old forms of intergovernmental relations in America. Distributive politics housed many of our urban poor, provided medical and social care for the disabled, helped Americans to move from their homes to work, and made possible continuing progress toward our national goals. Yet it has remained essentially dependent on energy released from the federal level. It has deployed personnel and dollars to state and local echelons but not necessarily problem solving capabilities. Creative federalism grows on the native strength of local institutions.

### Creative Conflict

The federal government is developing new relationships between itself and other decision-making centers, not only those in the state and local governments but between government and universities, voluntary and professional organizations, new public bodies and private business.

These partnerships in problem solving offer exciting possibilities, but no easy answers. There will always be tensions in these relationships, because the needs and interests of the partners are not identical. Indeed, creative federalism encourages the growth of institutions that will be independent of and in part antagonistic to the federal power. Nothing will be achieved if the partners become subservient arms of the central authority. Strain between Washington and the other centers is a necessary by-product of the decentralization, and autonomous decision-making is what creative federalism seeks.

This decentralization reinforces two of the stimulating tendencies of intergovernmental relations already noted—centralizing dependency and persuasion. It expands the scope of both vertical and horizontal operations.

48

## The Challenge of Creativity

These partnerships are the instruments of creative federalism but not its essence. America's present need is to find ways to gather clear, accurate information on the functions of society; to use this information and the understanding and wisdom which it can muster to consciously devise strategies, and program packages, which all change the boundaries and expand the options of public policy; and finally to try out and adjust these strategies before they are diffused and implemented generally. Thus, authority and sovereignty invoked by appeals to symbols are replaced by rational attacks on situations the American people find unsatisfactory or troublesome.

At rock bottom, this course of improving the flow of information, responding to it in imaginative ways, and refining the results in a limited setting before their general adoption is the process of creative federalism. If it is to be successful it must engage energies at the local and state as well as the national level, and among intellectuals, administrators, politicians, educators, businessmen, and technicians.

## The New Function of the National Government

The federal government has a special task in this process. Its role is not simply to provide channels and mechanisms whereby the national constituency can be played back against regional and local constituencies in the problem solving process.

This requires administrative inventiveness. But it is the best way by which the changing, transient environment can be made responsive to the needs of the men who inhabit it. It is the best way that American government in its broadest sense—as a congress of all the people—can successfully cope with the problem of urbanization.

# 4 DILEMMAS OF HOUSING POLICY

## Nathan Glazer

I was born in New York City, and live in Berkeley, California. Everyone knows about housing in New York. The slums of Harlem, Bedford-Stuyvesant, and the South Bronx are on the travel itinerary of every sophisticated tourist. Indeed, one might say the education of every American of some degree of social conscience hardly seems complete without some experience of the slums of New York. Many students go to New York after their graduation: some actually go to live in the slums, some work in the slums, and some live in buildings and areas that seem to have been taken over by the youth for their exploratory post-college year or years.

So almost everyone knows about the slums of New York. And not only the tourists and the students: the housing experts too have very often seen our problems in terms of the slums of New York—the old, poorly designed tenements, terribly crowded, too expensive, and badly maintained.

It was in New York where tenement house reform first began, at the turn of the century; New York where the first zoning law was passed; New York where the European workers' housing developments of the 1920's had their greatest influence and were imitated in a number of experimental projects; New York where public housing was born; and it is New York, and other crowded Eastern cities like it, that first come to our attention when we think of American housing problems.

And yet Berkeley is America too. It is a city of more than 100,000 people, continuous with the oldest and most densely populated parts of a large metropolitan area of 3 million. Its population is

25 per cent Negro, and it has its standard American urban proportion of welfare cases, juvenile delinquency, crime, and the like. But Berkeley has no slums, or hardly any. It has no tenements, that is, multiple dwelling units built specifically for the poor. The poorest people in Berkeley—and Berkeley is not a rich community—live in two-story motel-like apartment houses in pastel colors, with modern equipment; or in small stucco houses on little lawns, which they rent or buy. The most serious housing problem in Berkeley, according to the professionals, is some of the older large single-family homes that have been cut up into student apartments. It is on these that the city planners focus when they want to renew Berkeley. Of course when that happens, the students and the defenders of complex urbanity set up a howl, and fortunately the planners have not succeeded yet. There is no public housing in Berkeley. We have been talking about it for many years, and we certainly have people willing to live in the cheaper and better housing public housing makes possible, but it would be hard to find an appropriate site.

There are complaints about housing in Berkeley of course: professors and other middle-class people find that the nice houses are too expensive; graduate students find that cheaper houses are not too numerous and not too well maintained; students object to the dormitories in which some live, and find that the new low motel-like apartment houses in which more and more of them live are too expensive and don't keep out the noise; middle-class Negroes find it hard to buy the house they would want; and poorer Negroes find their housing too expensive. As in every American city, the proportion of Negroes is rising and the proportion of whites is falling, as the whites move beyond the hills to newer, more modern, and cheaper developments to the east. The proportion of Negro children in the public schools now stands at over 40 per cent.

As I have said, New York is not the United States; and neither is Berkeley. But Berkeley is in some degree typical of a good part of the United States—the West, the Southwest, and even parts of the Midwest—just as New York is in some degree typical of others. In New York, the poor—and by now, in our cities, that very often means the Negro poor—are crowded into tenements; in Berkeley they are rather less crowded and live in fairly good housing. In both cases their other problems are not very different; getting good jobs, staying off welfare, keeping the family together, getting a good education for the children, and keeping them out of the courts. And in between New York and Berkeley there are still other patterns. There are the great Midwest cities, from which there has been a huge migration of better-off whites, leaving a great deal of fairly good housing available for poorer Negroes. There is the South, where Negroes in the rural areas and in the small towns live crowded in ill-equipped shacks, and where their housing in the cities is the worst in the country, according to the census.

When one looks at this variety of American housing conditions, one is inevitably given pause in considering, just what *are* our housing problems, how serious are they, and what policies should we devise to improve them? One often hears that we need massive new housing programs in this country. No one disagrees. But that word "massive" must be taken with the proper understanding. If it is taken to mean that there is a massive shortage of good, decent housing in this country, and that vast quantities of housing must be built fast, on the basis of new and expanded legislation, then I would disagree. We need a great deal of new housing in this country; simply keeping up with growth and replacement means we should probably be building two million units a year, as against the little more than a million units we have built in each of the past two years. But the United States is not in the position of most countries, in which absolute shortages of housing exist. We find these shortages of course most prominently in rapidly urbanizing poor countries, in which people live in poorly built shacks without water and plumbing (which is what they mean in those countries by slums). But we find shortages too in fairly prosperous countries, such as Russia, England, France, West Germany, and even in the second most prosperous country in the world, Sweden. In all these countries one can legitimately say that there have been or are massive housing shortages, owing to such facts as the age of their housing, war destruction, the impact of war shortages on building, and rapid urbanization. Each of these countries has different housing policies; in each the problem of housing shortage has been or is so severe that the government makes some estimate of need, decides how much it can afford to build this year or the next, builds it, and in some way rations it out. There is an enormous difference, of course, between Russia and Sweden, in the quality of the housing, the size of units, the sophistication of the planning, the proportion of a tenant's income he pays for rent, and yet there is some similarity. In advanced countries with well-developed cities and industrial resources, this is the way things are.

They are not this way in this country. About 3 per cent of our housing stock is vacant, in good condition, and available for rent or sale. There is more good housing available for rent than for sale, more in the West than in the East, more in the country and in the small towns than in metropolitan areas, and yet, over-all, the United States offers a degree of housing plenty that is not to be matched in most other countries, whether poor or rich. Admittedly much of this housing is too expensive for the people who need it most; but if we look at certain measures of housing distress (for example, doubling up), we find that they are very low. Even among nonwhites, whose housing circumstances are poorest, only 5 per cent of families lived doubled up in 1960 (compared with 2 per cent for whites). These figures have almost certainly dropped further since 1960.

We have had massive housing shortages in the past, though none of them led to a consistent national policy to build and distribute enough housing to provide basic needs. After World War II we suffered from a severe housing shortage, since little housing had been built for fifteen years owing to the Depression and the war, ten million veterans were returning, and we had prosperity again. We still did not adopt a really national housing program, though we did pass a special act that permitted developers to make a great deal of money while building apartment houses. A great deal of housing was built, until newspapers discovered that builders were getting huge sums from the government to help overcome the housing shortage, and the program was stopped.

Our housing policy has been somewhat unique for an advanced industrial country, but one that is clearly in line with what most people want, and with what Congress is willing to support. The main point in our national housing policies consists of insuring mortgages for individual single-family houses through the Federal Housing Administration and the Veterans Administration, and setting a minimum standard for these houses—and thus in effect for most American housing—by determining what amenities must be built into such an insured house, how much space it must possess, what size plot it must be on, and the like. Then we leave it up to the builders and their customers and the banks to decide on the rest, such as how much to build, where to build, and at what mortgage rates. Actually only a small proportion of new single-family homes now have FHA or VA guaranteed mortgages, but FHA standards still determine what most new American housing is like. The local authorities play little role, except through the power of zoning and building code enforcement. But the character of the houses is set by the FHA requirements, the volume by general economic factors. FHA also insures apartment houses, but most are built on the basis of market demand without government aid.

We are concerned as a nation with the total volume of building, but less because we need the housing to overcome a general housing shortage than because homebuilding is an important industry, and national policy to maintain full employment makes it desirable for the homebuilding industry to operate at a high level. In 1963, 1964, and 1965 we built more than 1.5 million units each year. In 1966 we built only 1.2 million and in 1967 only a little more. In other countries, the amount of new housing is determined by national decisions: how much labor and materials are needed, how much they can afford, whether other needs are more pressing. In this country, we don't have anyone asking these questions— or if people are asking them, they are in no position to make the decisions that might be required by their answers. Interest rates have gone up recently, because of the war-induced inflation and because of the need to restrict the outflow of money to the high-interest countries of Europe; and as a result, homebuilding has declined. We are losing more than 300,000 units a year that might have been built, not because anyone decided we could do with 300,000 fewer houses this year, but because of general economic policies. Interestingly enough, that is just the figure that former President Johnson said he wanted to build for low-income and middle-income housing. If we could get the interest rates for new building down 1 per cent—and I am not enough of an expert to know how one does that or what its varied consequences on other parts of our economy and our international position would be—we would have that additional amount of housing and more, with no new policies. It would not be low-income and middle-income housing exclusively, of course, but the people moving into the new housing would release housing that would certainly improve to some degree the housing situation of poorer people.

The support of middle-income housing through FHA and VA is not our only housing policy, but it makes up most of it. In addition, we build some publicly owned (federally subsidized) low-cost housing, but very little—about 30,000 units a year. All told, public housing is only about 1 per cent of our housing stock, and 1 per cent of our people live in it. We also have an extensive urban renewal program, which eliminates a good deal of low-cost and poor housing and usually replaces it with expensive and better housing. Urban renewal is often called Negro removal because it so often takes place in the poorest parts of the city, with the poorest housing, and the poorest people—who are often Negro. A good deal of urban renewal in the past—and a good deal in the present—has destroyed cheap housing, forced people to move with inadequate compensation and inadequate help, required the poor to pay more for their housing, and has often been misguided as to how attractive the cleared and subsidized land would be to private builders. Thus the poor, in new and expensive housing, look bitterly at the empty lots on which they once lived, wondering why they were ejected.

Urban renewal has received a great deal of criticism, and some of the best studies of our urban problems in recent years have been powerful critiques of urban renewal. And yet the whole story is somewhat better than that. The poor move to better housing, on the average—though they pay more for it. They are getting more and more help and compensation when they move. Some of the housing that has replaced the old areas has been well designed, an aesthetic and financial lift to the city in which it is located. And perhaps most important, much of this new urban renewal housing is racially integrated; and in this country, where the whites so commonly keep out the blacks or move away when they come in, this is an achievement, of a kind of which we do not have many.

But compared with the FHA and VA housing, which has been built in vast quantities since World War II, all around our cities both public housing and urban renewal housing are pretty small programs. The FHA and VA house, with its government-insured mortgage, is just about within the reach of the family with a median income. The great argument in American housing is, what about the rest? Here is our great American housing policy that provides housing for everyone above an average income. We have a trickle of public housing for those below the average. But what are we doing for the great masses of the urban poor, and particularly the Negro poor? When we look at the slums of our great cities, and our small towns, we may well conclude that we are doing nothing, and that this is one of the greatest national disgraces in the United States. So foreign visitors conclude. Yet the matter is really not so simple. The conditions in the slums of our great cities are truly beneath the level the richest nation in the world should tolerate. But the fact is that some progress is being made in replacing these slums; in addition, many of the problems of these slums that we think of as housing problems are that only in part or hardly at all.

This progress in narrowing the spread of the slums is scarcely to be ascribed to any national, specific slum-clearance policy, for as I have said, both public housing and urban renewal are fairly small program. The slums are smaller because the American economy is dynamic. Recently, great quantities of new housing were built, a good deal of poor housing was demolished by highway building, urban renewal, and the like, and a good deal of poor housing too has been rehabilitated—because the economy has been prosperous and many people have been able to upgrade and improve their housing.

This improvement has been far more extensive than people imagine. Let us consider the most deprived part of the population in terms of housing, the nonwhites (92 per cent of whom are Negro—but our census figures only permit us to distinguish nonwhites for this comparison). Between 1950 and 1960, the proportion of nonwhite families in substandard housing dropped from 72 to 44 per cent; the proportion overcrowded (nonfarm only) dropped from 32 to 27 per cent. This improvement was much more moderate than for the white population—but it also occurred at a time of massive Negro migration to Northern and Western cities.

The improvement has continued in the 1960's. The proportion of nonwhites occupying substandard housing dropped between 1960 and 1966 from 44 per cent to 29 per cent. In cities of more than 50,000, the drop has been from 25 per cent in substandard housing to 16 per cent. Overcrowding is still severe and not declining rapidly: in 1966, 25 per cent of nonwhites lived in overcrowded housing, that is, more than one person to a room. The scale of these changes is really considerable. The number of substandard housing units occupied by nonwhites in 1960 was 2,263,000. By 1966 it had dropped by one-quarter, by 572,000, to only 1,691,000. The number of standard units in 1960 was 2,881,000. By 1966 it had risen 44 per cent, by 1,354,000 units. The decline in substandard units occupied by whites was even greater—from 6,211,000 units to 4,027,000 units.

One could argue that even our present-day housing policies, indifferent as they seem to the plight of the poor when we consider the magnitude of building specifically for the poor, may be sufficient to wipe out slums in a measurable time. Thus, Frank Kristof, research director of the New York City Housing and Development Board, concluded on the basis of a careful analysis of changes in New York between 1950 and 1960 that substandard and overcrowded housing in that city had been reduced during that decade —despite the great increase in Negro and Puerto Rican population, and the great increase in the low-income population—by about a quarter, and that it would be largely eliminated by 1980. Interestingly enough, another housing expert, William Grigsby, in an analysis of national housing trends in the 1950's, predicted the elimination of substandard housing by 1980, too. And this without any great new housing programs.

I have reviewed some of this material not to argue that we do not have a serious problem in the great slums and ghettos of

57

our cities—we do indeed, and a problem that all the intelligence and power of this nation may fail to overcome—but to argue that it is not primarily a *housing* problem. I will not use the occasion of a discussion on housing to argue in detail about just what kind of problem it is, but I would say we have three problems that look like housing problems, that are related to housing, that housing policies may affect—but are not housing problems. These three problems are problems of poverty, problems of neighborhood, and problems of segregation and ghettoization.

By saying we have a problem of poverty, I mean that higher incomes in the poverty groups would effectively solve their housing problems even without new housing policies. There are vacancies —which are too expensive. There is a housing industry capable of providing much more housing than it now does—if there were a market for it. Admittedly we would have to something about the abnormally high interest rates to make it possible for this industry to respond to new demands for new customers. The housing issue here is that the kind of housing that we in this country consider minimal—in terms of its construction, its amenities, its site, and so on—is likely to cost $20,000 a unit in our big cities, where the problem of poor housing is politically and socially most significant, and not much less in smaller places. Of course our poor cannot afford it, unless it is heavily subsidized, as in the case of public housing. It is not likely that it will be politically feasible to subsidize $20,000-a-unit housing to the point where all the poor live in *new* housing. But neither is it necessary that all the poor live in new housing.

To say the problem is poverty is a platitude and a truism— except for one implication. Once one accepts this, the question becomes, do we supplement the income of the poor, or do we subsidize housing so they can afford it?

The implications of these two approaches are vastly different. In one case the housing market responds to demand, in the other case the government commits itself to build and subsidize housing for the poor. There are varied arguments for one course or another. In the first case, there is the fear that increased income will not be used for housing, but for something less important to the poor family. (And we might answer, who are we to judge?)

58

There is the fear that increased income will only increase costs and profits to the builders and renters—but if there are sufficient manpower and materials available, this should not happen. There is the fear that it will be more difficult to encourage or require integration if new housing is built by the private sector. If we build public housing, we can make sure that our subsidies go to improve housing, and we can engage in policies that will contribute to integration. But there are drawbacks to the public approach too: the drawbacks of public action, which must inevitably be slow, surrounded by limitations and constraints, and which seems to arouse more hostility from those subject to it than private action does.

It is no simple matter to decide among the various ways of distributing the extra public funds that might aid the poor to get better housing; direct cash payments in the form of increased welfare, family allowances, negative income taxes, or what not; payments in the form of rent or housing vouchers; or subsidies of housing. And in the last case, we have the alternatives of subsidies to housing built directly by public authorities, or subsidies to homeowners of various kinds, or subsidies to business to encourage it to build and rehabilitate housing for the poor, or subsidies to nonprofit agencies to encourage them to build and manage cheap housing.

My own feeling—and this is why I shy away from the massive approach to our housing problems—is that many policies are desirable, and each has some benefits and some drawbacks. I think we need more in the way of direct payments to the poor; in some cases we might try the rent voucher approach, so the extra money is devoted to some need we as citizens think more important than some alternative uses. We already use this approach in the rent-supplement and leased housing programs. We certainly need public housing, and we should certainly try some new schemes to bring private money into housing for the poor and rehabilitation in the slums by government subsidy and tax incentives. We should try to expand homeownership among the poor, which many of them prefer, by subsidization of homeownership. What I am saying is that there is no simple answer. Suppose, for example, we decided we need five times as much public housing as we now build. In view of the political difficulties of finding sites for new housing, of trying to prevent ghettoization in public housing, of a certain amount of public resistance to public housing, I doubt that we could get anywhere near this goal. Perhaps we could build twice as much as we now do. And even that may be too much to expect.

We heard a good deal about rent supplements and leased housing. These are excellent schemes for distributing the poor among the less poor, preventing the institutionalization of housing by means of big projects, trying to combat segregation and ghettoization. More money should be appropriated for them. Yet it would be naive to see this program encompassing hundreds of thousands of families—at least within the next five years or so—even if we had the money. The not-so-poor are just not so happy about having the poor among them, the whites at having the Negroes among them, landlords at taking in tenants who they fear might not keep up the property as well, and the like. This is inevitably a kind of tailored and handwork operation: you have to find the apartments and the houses, find tenants suitable to the landlord, execute special contracts, and so forth. Let's not expect too much of it—too much in *scale*, that is.

I supported the efforts of Senator Robert Kennedy and Senator Charles Percy to bring more private capital into the building and rehabilitation of low-cost houses. I suspect the scale of these approaches will not be too great. There is simply not that much profit in providing housing for the poor; and if we tried to provide the kind of subsidy or incentive that would make it profitable, we would have the same kind of howl and scandal that we had when we tried to bribe builders into putting up housing fast after the war. In addition, building and managing housing for the poor is simply not a clean business. The poor have problems —obviously. Who has managed housing for them in the past? Very often slumlords, often drawn from the same group whom they were housing. They knew their tenants—knew them well enough, that is, to make a profit. And they had to be pretty tough and hardhearted to do it, and in view of the work they put in, the profit was not very big either. For some reason I don't quite see this as a business for Ford or General Motors. Why should they want to get into it? After all, the slumlord's son rarely followed in his father's footsteps, if he had alternatives.

One of the reasons, of course, for bringing big business into rehabilitating and building houses for the poor is the hope that their technical expertise and capacity for handling complex problems will help reduce that overwhelming cost of $20,000 for a decent housing unit in a big city. Certainly we should try this approach—as we should do more to deal with restrictive building codes and labor union practices. But once again I don't put too much hope in a massive technological breakthrough. The reason is that those countries that don't have quite the same problem of restrictive building codes and labor union practices and that do use all the available technical skill in building still are not able to cut the costs of housing that much. Housing is expensive everywhere—in Russia, in Poland, in Sweden.

The second problem that is somewhat intermingled with that of housing and that is yet distinct from it is that of neighborhood. When people are shocked at American slums, it is often not the quality of the housing as such that creates the shock. It is rather the level of maintenance of public facilities as well as of housing; it is the absence of public facilities (parks, open space); it is the visible presence of social problems—idle men lounging on street corners, men drinking in hallways, women clearly drunk or unbalanced wandering down the sidewalk, and so on. Better housing will not necessarily solve these problems. If maintenance is poor, it is probably because of poverty, individual and public, as well as for other complex reasons—city indifference, differences in background and expectations, absentee landlords, and so on. If there are insufficient public facilities and amenities, and the streets are filled with idle children, once again this is not only a housing problem. If many people suffer from one kind of problem or another, once again, better housing is not likely to make a decisive contribution. It is for this reason that, while public housing is often preferred to slum housing, the public housing tenants are often not that much happier with their lot. The maintenance is better, there are somewhat more public facilities—though generally not enough—but the social problems have not gone away, except insofar as admissions policies keep out drug addicts, criminals, and the like. Many studies show that people in the slums generally object to the neighbors and the neighborhood as much or more than they do to the housing, and the people in public housing projects have the same objections.

How does one deal with the problem of neighborhood quality? That is much harder than improving housing. If everyone works, that improves neighborhood quality. If poverty is mitigated by greater cash income, that may improve it. If cities can spend more on keeping up their poor areas, that will help. If we find ways of dealing with those intransigent problems of social pathology that are so in evidence in the slums of American cities—crime, drug addiction, alcoholism, family instability—that will help too. But once we detail what is needed to improve neighborhood quality, it is clear that we will not have any quick and easy solutions.

The Model Cities Program offers some help here. I think it promises too much, requires achievement in too many areas, and calls for too much planning and coordination, which is very time-consuming. But the more intelligent mayors see that they can use this money to do one thing in the slums, and this is clean them up. They can improve garbage collection, rebuild streets, add playgrounds and neighborhood centers, improve code enforcement and maintenance. The Model Cities Program will not solve some of the more difficult social problems, but it does permit the hard-pressed city government to do those things that can be done and only require more money.

Finally, a third problem we really have in mind when we talk about housing is segregation. Negroes live together not only because they are poor (they are), and not only because many want to (many do), but also because they are discriminated against. The factors leading to residential segregation are far more difficult to deal with than poor housing, though the two are closely related. Discrimination means Negroes are forced to take the poorest housing, and pay more for it, at every income level; the forced segregation contributes to the atmosphere of despair, hopelessness, and anger that helps make the ghetto a miserable place, whatever the technical quality of the housing.

But what does one do about segregation? One can pass laws, and many states and cities have them. They don't seem to help much in actually affecting patterns of segregation, though they do help in individual cases, and they do put on record a public and official commitment against segregation. We now have federal regulations against discrimination in all publicly supported housing; once again, the impact, aside from some public housing and some urban renewal housing, is not great. We don't see many integrated suburban tracts. There are no massive public policies I can conceive of in a free market democracy that will ensure the rapid breakdown of racial residential segregation. In a democracy, people can resist the passage of laws and the strong implementation of policies to increase integration. In a free market economy, many people will exercise their right to move away, making the integrated neighborhood a transitional one between all-white and all-black. And yet there is no question in my mind that residential segregation is a far more serious threat to American democracy and the health of American society than poor housing. To face a most serious problem, one that hurts both blacks and whites, that damages the quality of education, that encourages racial polarization and extremism, and to see no immediate solution that will be acceptable to the majority of the American people is to be in a bad situation indeed.

But there are certain things we can do, and there are certain measures that are still politically feasible. Our greatest strength in dealing with this problem is the commitment of the presidency and the executive establishment to racial integration. This commitment can be made stronger, and exercised more effectively in the oversight of public housing programs, urban renewal programs, and rent-supplement programs, FHA and VA developments, and in all the other areas in which federal oversight and regulations play a major role. In many cases there will be a tragic choice between more and segregated housing, and less and integrated housing. I have no advice on tragic choices; men of good will will have to decide in each case which is better.

There is one large area of policy that indirectly works for more integration and does not raise the politically sensitive and difficult problems of outlawing segregation and promoting integration directly. That is raising the economic status of the Negro. The better-off Negro does have more choices, more opportunity, more possibility of locating in a white or integrated neighborhood, if he desires to. I would say the political and the economic arms of policy can work together on this problem. Both are necessary— but more of one can make up for less of the other. I have no doubt that if Negroes were magically and overnight to achieve the same economic level as whites, the problem of segregation would be enormously reduced. It will be a long time before that happens; in the meantime, various partial and ingenious approaches are possible, and all should be expanded.

In the end, I would conclude there is no one massive housing problem in this country; there are many problems, some of which involve strictly better housing, most of which involve housing and other things. Nor do we find overwhelming housing shortages, or overwhelming housing deficiencies, whatever the evidence of the great city slums. We need larger and better housing programs— but not so much larger or so very much different that they are outside the bounds of political possibility. In 1966, there were only about 3 million substandard housing units in metropolitan areas. Some of the people living in those units are probably for some reason happy with their housing, and others in officially standard units are not. If we were to decide to supplement the housing costs of the poorest 3 million urban families, in addition to the poorest 700,000 we now subsidize, through a mixture of public housing, rent supplements, incentives to private business to build and rehabilitate, and a variety of other mechanisms that are already in existence or have been proposed, we would certainly have solved the worst of our housing problems. The cost in annual subsidies would be of the order of $1.5 billion.

This is the largest program that I can envisage as being justified, and even so it is rather smaller than many that have been proposed. We presently add about 50,000 units a year, under our various programs, to the stock of housing that is subsidized for the poor and low-income groups. I can envisage tripling or quadrupling that number, by expanding the old programs, and adding new ones along the lines of the Kennedy and Percy proposals. That would certainly speed up the rate at which we improve our housing, and we might find, long before we had reached the 3 million figure, that our poor housing problem had been reduced to very small dimensions. But this would not solve our urban problems. We would still have poverty, poor neighborhoods, and residential segregation, and these will be much more difficult to deal with. At a cost of adding a billion or a billion and a half dollars a year to the federal budget, we would have a situation in which no poor family in our big cities would need to live in substandard housing. I do not underestimate the difficulties of getting Congress to appropriate even such a sum—but we are not now talking of tens of billions. A billion or two would not loom very large in our federal budget, and would permit us to face the world and ourselves with less embarrassment.

# 5 URBAN TRANSPORTATION

John R. Meyer

The popular view when discussing urban transportation in American cities today is to decry its sorry state. Newspapers and journals are filled with talk of an "urban transportation crisis," of the "difficulties of getting from here to there," and so on at great length. Matters are reported to be getting worse—and very quickly. Everyone has his own favorite traumatic experience to report: of the occasion when many of the switches froze on New York's commuter railroads; of the sneak snowstorm in Boston that converted thirty-minute commuter trips into seven hour ordeals; of the New York transit strike of January 1966 that converted Manhattan Island into a traffic nightmare; of the extreme difficulties in Chicago and other Midwestern cities when some particularly heavy and successive snowstorms were endured.

There is, of course, always some danger in generalizing on the basis of casual observation or the isolated experience. Urban transportation is no exception. Thus, when one compares actual figures on the performance of urban transportation systems in the United States today and yesterday, one generally discovers some slow but steady improvement. The numbers are few and scanty but nevertheless systematically point toward a better situation now than previously. Data on the time required to travel between two points in American cities during rush hours today and, say, ten or fifteen years ago almost invariably suggest that the time required to make a trip has decreased slightly, usually in the vicinity of 10 to 20 per cent. Or, alternatively, the average highway vehicle speed during rush hours in the central parts of American cities has slowly but surely climbed upward over the last ten years from approximately 15 to 25 miles an hour to somewhere in the

66

vicinity of 25 to 30 miles per hour or slightly more. Needless to say, the figures vary widely from one set of circumstances to another. Moreover, there are some isolated but important cases where there has been deterioration, either in service levels or in actual performance.

That some improvement has occurred is hardly surprising. Americans have been spending a good deal of money recently to improve their urban transportation systems. First, they have made large private investments in automobiles: since 1945 the number of passenger cars registered in the United States has increased from just under 26 million to well over 70 million. Billions of dollars have also been invested by federal, state, and local governments in new highways, the current rate of expenditure being about 3 billion dollars per year on urban highways alone. Transit systems have also been improved. For example, the total number of buses registered in the U.S. has increased from approximately 112,000 in 1945 to over 150,000 today. Some extensions and improvements have also been made in rail mass transit systems. About the only major deterioration in urban transportation facilities in American cities in the postwar period has been of rail commuter trains, which provide high-performance express services between central business districts and suburban communities.

Improved speeds should not, however, be confused with any shortening of the time required to perform a typical commuter trip in American cities. Actually, on the basis of the limited data available, the time required for the typical or average commuter trip seems to have remained more or less constant. The explanation, of course, is that the distance of the average commuter trip is longer today than it was a few years ago. This lengthening of commuter trips reflects an increasing tendency of Americans to live relatively more in single-family dwellings in the suburbs and less in multiple-family or apartment house dwellings in the central city. This choice of single-family dwellings means that Americans live at lower residential densities and, as long as everything else remains equal, must commute a greater distance to reach their jobs.

Actually, not everything else has remained equal. In particular, workplaces have also dispersed to a considerable extent in American cities On the whole, this dispersal has meant a removal of jobs from central business districts or urban core areas to the suburbs. This dispersal of jobs toward suburban locations has tended to offset some effects of the residential dispersal, making it possible for many workers to find the kind of housing they seek while living closer to their jobs. Nevertheless, these locational trends, dispersal of workplaces and jobs, have not been completely offsetting. The typical trip to work is clearly of somewhat greater distance in the American city now than ten years or so ago and just sufficiently longer, seemingly, to offset whatever slight improvement has been made in the performance speed of urban commuter transport.

One reason, then, for the talk of an urban transportation crisis in the United States today perhaps lies in a failure to meet anticipations. Many commuters expected to reduce their commuting times as systems improved, but instead found themselves barely able to maintain the status quo in terms of time requirements. Another reason for talk of crisis, almost certainly, is that the rate of improvement in the performance of urban transportation systems during rush hours has been markedly inferior to that experienced during off-peak hours. Specifically, the ability to move quickly about American cities during nonrush hours has improved in a truly phenomenal fashion. Using limited-access urban expressways, or freeways as they are called in the Western part of the United States, one can easily achieve speeds of 55 miles an hour or more.

It is, of course, an exhilarating experience to move through a congested urban area at such speed, with relative safety and comfort. The contrast between these high-performance speeds during nonrush hours and the relatively slow 25 to 30 miles an hour experienced during peak commuter periods can be disheartening. Quite humanly, the commuter would like to duplicate his off-peak speeds and experiences. The difficulty is that to accommodate these wishes, particularly if most commuters decide to go individually in private automobiles, would require a truly remarkable amount of highway capacity. While there has been a good deal of loose talk about how paved-over American cities are or have become, accommodation of most commuter needs by private automobile at high speed would require new expressway construction in or near central business districts that would pale previous efforts.

Otherwise, the automobile and the urban expressway generally economize on land requirements for transport purposes in urban areas. A common fallacy is to believe that reliance upon private automobiles as the basic mode of urban transportation necessarily results in a higher percentage of urban land being allocated to street and highway use.

Perspective can be obtained on land use requirements for transportation by looking at the entire urban area, including residential neighborhoods. In the over-all picture, the major land needs for highways and streets are attributable to local streets. Moreover, good local street access is required for police and fire protection, sanitation, local deliveries, and so forth, however one decides to organize the longer distance commuter transportation system. Local streets can require from 15 per cent or so of total land in a residential area to as much as 30 to 35 per cent. In American cities the lower percentages for local streets have been achieved in newer residential areas built on the premise that most households would have an automobile to serve their transportation requirements. Under such circumstances, long rectangular blocks can be used which economize on the total amount of land allocated to streets. By contrast, in the denser neighborhoods of older cities, where it is expected that most people must walk to and use public transportation, smaller, square blocks are typical. In these neighborhoods streets can account for as much as 30 or even 40 per cent of total land use. Of course, residential densities will also be higher in these communities as a rule, so that these older cities remain relatively more compact than the newer, highway-oriented cities.

Major arterial highways, freeways, and expressways generally require relatively small amounts of land. Even in Los Angeles, which has what is probably the most ambitious and fully developed expressway system in the United States, the total space required for the freeway system when it is completed will be only about 2 to 3 per cent of the entire land area of the city. One can dispute and debate about the specific amounts of land required for expressway systems, but rarely should the total amount of land for high-performance highway systems exceed three per cent of total land or about one-sixth or one-seventh of the land required for local access streets. The total land requirement for all kinds of streets and highways in urban areas, therefore, is likely to run between 20 and 35 per cent of total available land. Again, the lower percentage is far more likely to be achieved where the street design is specifically laid out on the premise of widespread auto ownership. Not only can more efficient block layouts be achieved under such a premise, but it should also be noted that high-performance expressways or freeways tend to move many more cars per hour per unit of space than lower performance facilities. That is, specialization of highway facilities achieves economies in land use.

There are other ways in which highways and highway vehicles represent an efficient adaptation to American circumstances. Specifically, under United States cost conditions, it is rather difficult to design a good public transit service at low cost except where very large volumes of people move along a common corridor. At corridor volumes of approximately 5,000 to 7,000 per hour or less and equivalent levels of service, public transit costs will be approximately equal to private automobile costs. Even then, only a bus transit system would be cost competitive with the automobile at such volumes. Generally speaking, a rail transit system becomes as low in cost as a bus system only at volumes of 18,000 to 20,000 people or more per rush hour along a corridor. Only New York and Chicago have transit corridors that record volumes much in excess of 25,000 per hour, and only Boston and Philadelphia have corridors that approach the 20,000 to 25,000 volume range per hour. With dispersal of jobs and residences, of course, there are many commuter trips increasingly performed along corridors with very low volumes, well under the 5,000 or so required to equalize public transit and automobile costs of equivalent services.

Private automobile transport in the United States derives some advantages from labor and capital costs, as well as from reduced residential and workplace densities. Specifically, automobile commuting is essentially a "do-it-yourself" activity. It requires little or no hired operating labor. The public transit system, on the other hand, usually does need hired operating labor. Therefore, it is rather in the position of the domestic servant in American society competing against washing machines, electric dishwashers, vacuum cleaners, and other consumer durables. The automobile, in short, is in keeping with the American pattern of responding to ever higher labor costs by substituting one's own labor and capital (in the form of consumer durables) for hired labor.

While the automobile, complemented by public transit buses, is probably a fairly efficient accommodation to urban transportation requirements in many if not most American cities, it carries with it many dislocations and traumas. In particular, efficient use of automobiles and other highway vehicles in many older American cities requires that large high-performance highways be built right into and through already heavily populated central areas. This, of course, means that many people and workplaces must be displaced to make room for the new highway construction. Though these new highways may require only a small percentage of total space in the city, this does not lessen the impact on those who are forced to leave their homes and relocate their businesses because of the highway construction. Thus, though the construction of these new highway facilities may be broadly justified by the gains to society as a whole, it is quite clear that the incidence of cost and trauma on those individuals who are displaced can be very high indeed. Unfortunately, American society has not yet developed adequate mechanisms for compensating these individuals.

Often, other social costs are also incurred because of new highway construction. For example, highway engineers have found that it sometimes reduces local community objections if they can find routes that minimize the number of houses that must be taken for construction of new highways. Furthermore, the total cost of the whole enterprise can sometimes be reduced if one can keep the highway in open and unoccupied land that is already in the public domain. An unoccupied urban area in the public domain well describes a public park, and so there has been a certain attraction on the part of highway builders for routing new urban expressways through and around the fringes of park lands. The difficulty, of course, is that reduction of available park and playground space tends to reduce the amenities of urban living in important ways. Thus, if the highway builder attempts to avoid household and workplace displacement by staying away from built-up congested areas, he in turn very often imposes a new kind of social cost on the society.

Sometimes, too, highway builders have shown a remarkable

insensitivity to other urban needs and requirements. For example, they have been known to build urban expressways and interchanges to the same specifications used in rural areas. This usually means that a great deal of land is consumed for interchanges, median strips, and other highway embellishments which on close calculation might not be strictly necessary or even terribly functional in urban areas. When one considers the additional fact that most urban trips are relatively short, well under ten miles, it is not entirely clear that very much is gained by designing urban highways to sustain peak-performance speeds of 60 or even 70 miles per hour as opposed to only 40 or 50 miles per hour.

New expressways may often be rather poorly located as well. For example, some new facilities have divided what were previously relatively homogeneous and well articulated communities. A badly placed highway can thus disrupt community life in an unfortunate way. On the other hand, the construction of new high-performance highway facilities also drains traffic off neighboring parallel streets, so that some offset is achieved in the sense that the older streets become safer and less traversed.

Still another difficulty is that under certain climatic and topological conditions, (such as those found in Los Angeles, Salt Lake City, and a few other American cities) highway vehicles can create a most offensive and dangerous smog. Indeed, smog nuisances have been created even under relatively conventional North American climatic and topological conditions. Smog control devices and programs to reduce these hazards are therefore under development. Some considerable progress has been made, but there is still much to be done. It should be observed that Europeans may never encounter quite these same difficulties, since European cars tend to be smaller and burn less fuel; smog created by operation of a vehicle is not strictly proportional to horsepower, but reduction of total horsepower does alleviate the problem.

Safety and noise problems can also be created by automobiles and new urban expressways. Here, however, the picture is somewhat more complex. For example, other things being equal, high-performance expressways reduce pedestrian deaths caused by automobiles in urban areas. In general, high-performance limited-access expressways tend to have a relatively good safety record in terms of accidents per million miles of travel when compared with conventional streets or highways. On the other hand, high-performance expressways are likely to increase the total miles traveled and the speed at which those miles are traveled. Similarly, there are pluses and minuses on the noise problem attached to use of automobiles or trucks in place of horsedrawn or other vehicles or the use of high-performance expressways in lieu of conventional streets and highways.

The preceding discussion has obviously concentrated on the highway and the automobile as major modes of urban transportation in the United States today. Such a concentration is easy to justify. About 64 per cent of American commuter or work trips are made in private automobiles. Another 14 per cent or so are performed entirely by buses on highways. Rail transit and walking account for only about 10 per cent each. For other trips, that is for shopping, recreation, and personal business, the automobile is even more dominant. Thus, as even the casual observer of the American urban scene quickly recognizes, the automobile is certainly a fixture, at least for the moment.

And as the previous discussion suggests, the reasons are not hard to find. The automobile, though a noisy, smelly, and somewhat unsafe monster, is also a highly efficient transportation mode given the lower residential densities, increasingly dispersed workplaces, higher labor costs, and relatively low capital costs that characterize the American scene. The real question about urban transportation in the United States today is whether we will have the good sense, self-discipline, and social organization to mitigate the many bad effects of the automobile while still retaining its many advantages.

# 6 POSTWAR CHANGES IN LAND USE IN THE AMERICAN CITY

John F. Kain

United States metropolitan development since World War II has attracted a great deal of attention, both here and abroad. The decay of central cities and urban sprawl, that is, the extensive low-density development of suburban America, have been the subject of considerable comment, much of it uncomplimentary. This pattern of central city decline and suburban growth is the result of three intimately related forces: first, rapid dispersal of employment from dense central cities; second, extensive metropolitan growth associated with low-density residential development; and third, the accelerated growth of massive Negro ghettos in the nation's largest metropolitan areas.

The present spatial structure of United States urban areas and the nature of most so-called "urban problems" are caused by the interaction of these three factors. The first and second have close parallels in most other nations, particularly those that are approaching United States levels of economic development. The third, in its most specific form at least, is unique to the United States. My objective here is to analyze these three factors and their effects on American metropolitan development. The discussion especially focuses on the impact of the Negro ghetto.

The location, or spatial distribution, of employment is the most important determinant of urban form or structure. Changes in the location of industry are the principal determinants of changes in urban form.

Since World War II employment dispersal within American metropolitan areas has been very great. What began as a relative decline in the growth of employment in central cities became for an increasing number of metropolitan areas an absolute decline. While the evidence suggests that all urban areas experienced employment dispersal following World War II, for statistical reasons these redistributions are most easily demonstrated for the forty largest metropolitan areas. For these forty metropolitan areas statistics are available giving central city and suburban employment in manufacturing, wholesaling, retailing, and selected services during several postwar years. During the first decade following World War II (1948-1958), wholesaling employment declined in about one-third, manufacturing employment in about one-half, and retailing employment in about three-fourths of the central cities of these large metropolitan areas.

These averages, of course, hide a great deal of variation. Large increases in employment took place in most of the spacious, low-density central cities of rapidly growing Southern, Western, and Southwestern metropolitan areas. The explanation for these increases in central city employment is, of course, the favorable confluence of high metropolitan growth rates and large tracts of unencumbered vacant land within city boundaries. Similarly, the older, dense manufacturing cities of the East and Midwest experienced especially large declines. Figures for the most recent period (1958-1963) suggest an acceleration, if anything, of the postwar trends toward dispersal. For example, the central cities of the same forty metropolitan areas lost an average of over 17,000 manufacturing jobs in only five years.

These trends in industry location are the result of basic changes in production and transportation technologies. Intercity and intracity motor trucks freed most producers from having to crowd into the limited area near deep-water ports or railroad marshalling yards. Only firms using very large amounts of bulk commodities remained closely tied to the ports and railways. For an increasing number of firms, outlying locations near major intercity highways and suburban beltways became more advantageous. When their centrally located plants wore out, many of these shifted to new locations in outlying areas.

The principal effect of these changes in freight and communications technologies was to make locations throughout metropolitan areas more uniform in terms of transport cost, reducing greatly the former locational advantages of central areas conferred by the concentration of freight and passenger terminals there. The locational impact of these changes was reinforced by other factors, including the adoption of space, extensive methods of production, materials handling, and sales epitomized by the spacious single-story warehouse and factory, the forklift, and the supermarket. All took their toll on the former competitive advantage of crowded and dense central areas. The fact that older central areas, which were developed during the era of the steam locomotive and horsedrawn wagon, frequently had street layouts and circulation patterns unsuitable for these new technologies aggravated their condition. The historical location of a number of clusters of activities for which externalities remained important and the cheap space resulting from sunk investment were almost the only advantages retained by central areas. In the absence of these historical factors, the process of employment dispersal might have been still more rapid, a conclusion which is consistent with the experience of the newer metropolitan areas of the West and Southwest United States.

This interpretation of postwar metropolitan development differs substantially from that commonly advanced by commentators both in this country and overseas. Many of these individuals, finding the prevalent low-density form of metropolitan development repugnant for one reason or another, have pointed an accusing finger at the rapid growth in automobile ownership and have called for severe restrictions on the ownership and use of private cars in order to prevent the complete destruction of American cities. Nor is rhetoric of this kind limited to the United States. Many other countries are simultaneously experiencing suburbanization of employment and population and rising car ownership. Invariably this leads to the conclusion that the increase in car ownership is the cause of the lower-density development. It is but a short step from this conclusion to the view that ownership and use of private automobiles should be banned or at least severely limited.

Yet if changes in the location of jobs are regarded as the principal determinant of postwar changes in metropolitan structure, the rapid growth of car ownership and use would appear to be only one, and perhaps a minor, factor. Viewed from this framework, changes in intercity and intracity freight transport appear to be far more important and fundamental causes of dispersal.

The automobile's contribution to this aspect of dispersal would appear limited to easing the difficulties of assembling a labor force at more remote locations. Also, insofar as the widespread ownership of private automobiles is partially responsible for a more dispersed pattern of residential development, it may have affected in this way the location of retailing and other service employment closely linked to these residences. The internal combustion engine may have been one of the major technological causes of the postwar pattern of low-density development, but the motortruck would appear to be much more important than the motorcar.

The automobile's influence on the suburbanization of metropolitan populations and the trend toward low-density residential development may have been somewhat greater than an industry location. Even so it must be regarded as simply one of a large number of forces responsible for these changes. Moreover, research indicates that the postwar pattern of residential development is as much, or possibly even more, a cause of the rapid growth of car ownership as the converse. Rapid increases in per capita incomes reinforced by the pent-up demands and large savings accumulated during World War II appear to be more fundamental factors underlying both. Cheap credit, favorable mortgage loan terms, accumulations of savings, rapid family formation, the postwar baby boom, favorable tax treatment, a strong preference for home ownership, and the suburbanization of an ever larger number of jobs must all be regarded as important causes of the suburban boom. The private automobile certainly abetted the process by providing an ubiquitous and reasonably cheap form of transport to and from the now dispersed residences, workplaces, and other activities. However, other forms of commuter transport (for example, walking or an extensive bus system) could have served this more dispersed pattern of workplaces and residences nearly as well. Widespread ownership of private automobiles hardly can be regarded as the sole or even the principal cause of the low-density pattern of postwar residential development in the United States.

Some of these factors, in particular the favorable tax treatment of home owners, the specific mortgage institutions, and consumer preferences, may be peculiar to the United States. However, increases in family incomes, employment dispersal, and the availability of ubiquitous and relatively cheap forms of commuter transportation are common to many countries. Thus, it would appear that continued declines in the density of residential development might well be anticipated in other countries. This conclusion is consistent with both my personal observations of the patterns of residential development in other countries and the impressions I have obtained from secondary sources.

Large-scale migration from the agricultural South to large metropolitan areas is one of two basic mechanisms responsible for the rapid growth of massive Negro ghettos in the largest United States metropolitan areas. Housing market segregation is the other. At the turn of the century, nine out of ten American Negroes lived in the South, mostly in rural areas. The corresponding proportion of the white population was only four out of ten. The rapid expansion of demand for labor by Northern industry during World War I prompted the first large flow of Negro migrants from the South to Northern metropolitan areas. The flow continued thereafter, abating somewhat during the Depression, accelerating greatly during World War II, and continuing at a high level to the present. As a result of these large-scale migrations, the proportion of all Negroes living in the North and West rose from 10 to 40 per cent between 1900 and 1960.

These large-scale movements of poorly equipped Negro migrants from the rural South would have created substantial assimilation problems even if they had been evenly distributed geographically within the destination regions. This was not the case. Ignorance, low incomes, and most importantly discrimination confined them, and more seriously their children and their children's children, to geographically limited portions of major metropolitan areas. The proportion of Negroes residing in places classified as metropolitan in 1960 increased from 27 per cent in 1900 to 65 per cent in 1960. In 1960 the 24 metropolitan areas with one million or more persons contained 38 per cent of the total Negro population, and their central cities, 37 per cent. Comparable figures for the white population are 34 and 15 per cent.

The initial settlement of Southern Negro migrants in central city ghettos was not unusual in itself. In this respect Negro migrants shared the experience of earlier white immigrant groups. However, their experience diverged in two important respects from that of the residents of European immigrant ghettos in American cities. They were more highly segregated than those European migrants, and their residential segregation, in general, has increased rather than decreased over time. The result of rapid Negro immigration and housing segregation has been the development of massive Negro ghettos in large metropolitan areas.

The growth of the massive central city ghettos has been particularly rapid since the start of World War II. Between 1940 and 1960, the total population of all metropolitan areas increased by 33 million whites and 6.4 million Negroes. Eighty-four per cent of the Negro increase occurred in the central cities, while 80 per cent of the white increase occurred in the suburbs.

An even sharper contrast appears in the 24 largest metropolitan areas—areas containing more than half the total United States urban population in 1960. Between 1940 and 1960 their white populations increased by 16 million and their Negro populations by 4.2 million. Only two-tenths of 1 per cent of this net increase in the white populations occurred in the central city, as compared to 83 per cent of the Negro increase. The result has been a rapid growth of massive central city ghettos and a significant change in the racial composition of central cities. During the decade 1950-1960 the Negro population of these 24 central cities increased by 2 million. The resulting growth of central city ghettos displaced large numbers of whites; these same cities lost more than 1.5 million whites during the decade, a process that continues to the present.

The significance of these data is even more obvious for particular metropolitan areas. For example, between 1950 and 1960, Cleveland's central city gained 103,000 Negroes and lost 142,000 whites, while its suburbs gained 367,000 whites, but only 2,000 Negroes. There were only 6,000 Negroes residing in all of Cleveland's suburbs in 1960, as contrasted with over 900,000 whites.

While the most serious consequences of housing market segregation are borne by the Negro community, it would be incorrect to conclude that Negroes bear all of the costs. There are substantial costs imposed on various groups of whites, and the aggregate social costs are larger still. Billions of dollars of federal subsidy have been expended in central cities in an effort to halt their deterioration. These programs have been notable principally for their lack of success. There are many reasons, but housing segregation and the rapid growth of the ghetto must be numbered among the most important.

Residential urban renewal programs are intended to modify the mix of central city residents and housing in favor of high incomes and high-quality housing. Growth of the central city ghettos subverts this objective in at least three ways. First, low-income households demand low-quality, and frequently substandard, housing. The enforced concentration of low-income Negroes creates massive slums in central cities. Second, the entrapped Negro demand maintains central city land values at high levels, inhibiting either the renovation or the clearance of central areas. Given these costs, high-quality housing can be created more cheaply in suburban areas. Enough more cheaply it would appear, to pay large commuting costs for many centrally employed white workers. Finally, a large and rapidly growing, low-income, Negro population makes central residential areas less attractive to whites. Part of this may be attributed to racial prejudice, but there are other reasons as well. For example, schools and other public services in, or near, such low-income neighborhoods are likely to be inferior to those provided in high-income neighborhoods and communities.

Federal policy in general has attempted to improve the conditions of central city neighborhoods by means of cash subsidies to acquire or renovate properties inflated in value by housing market segregation. These programs have not been successful and the quality of central city neighborhoods has continued to decline, at least relative to other parts of metropolitan areas. Clearance of a Negro slum typically is followed by its reappearance a few blocks away and, with the continued growth of the Negro population, central cities increasingly have become lower-class slums. These outcomes are the product of a housing market grossly distorted by racial discrimination. If the ghetto could be eliminated, the prospects for central city renewal by public or private means would be much improved. The result might well be even smaller central city populations; but the quality of central city housing and environment would improve markedly.

Centrally located housing in countries not afflicted by the pattern of racial segregation prevalent in the United States should benefit from these same forces. Employment dispersal should reduce the demand for centrally located housing by low-income workers, thereby encouraging both private and public redevelopment of many of these areas to lower density and higher quality housing. From casual observations, I would conclude that this is occurring in a number of the world's cities where the housing market is functioning to renew central residence areas, unlike American cities.

It is implicit in the above that racial segregation aggravates the problem of providing urban transportation services. As the central ghetto continues its inexorable growth, centrally employed whites are pushed further and further from their places of work. This leads to demands for improved, higher speed and subsidized long-distance commuter transportation. Passing these centrally employed white commuters daily on their way to work are ever increasing numbers of ghetto Negroes, who find it necessary to travel even longer distances to reach suburban jobs.

Were it not for housing segregation, a much different commuting pattern might exist. Many present white residents of distant suburbs might live in central areas near their places of work. Similarly, many present Negro residents of the ghetto would reside in outlying neighborhoods near their suburban work places. These changes in residence location would greatly reduce the amount of reverse commuting and crosshauling, shorten trips to work, and lessen the demand for expensive, difficult to provide, and specialized commuter facilities. Continued dispersal of industry might lead to a further reduction in the length of the journey to work. These improvements may be more quickly realized in countries that are not subject to these forms of discrimination.

While my knowledge of metropolitan development in other nations is scanty, it would appear that the forces of employment dispersal and suburbanization are at work in other developed countries. Moreover, it seems possible that new population centers of less developed countries may never acquire the compact and highly concentrated employment pattern characteristic of the large metropolitan areas of developed nations during the nineteenth century. Such generalizations are based on economic and technological factors common to both the United States and other countries. However, the Negro ghetto is peculiarly American. Therefore, it might legitimately be asked if there are any reasons, in addition to intellectual curiosity, for persons in other countries to be concerned with its impact. This question must be answered in the affirmative.

First, nations with large less-privileged racial or ethnic minorities might well take American experience as a warning of the costs that can result from housing market discrimination. This would be particularly true where the minority is as easily identified as the American Negro.

Second, nations with no large and easily identifiable minorities should be interested in the effect of racial segregation on the patterns of American urban development for different and more subtle reasons. American experience is frequently used by metropolitan planners, social scientists, and others to forecast conditions in other countries at some future time when they reach, or approach, United States levels of affluence. Projections of this kind can be of considerable value, but they must be made and interpreted carefully. Important differences between American conditions and those of other countries must be identified and their effects evaluated before any conclusions are drawn from American experience. Residential segregation of Negroes may be one of the most important of these differences. I believe that racial segregation and the rapid growth of the ghetto have affected strongly the operation of the housing market and hence the development of American metropolitan areas.

NOTE: The research on which this paper is based was supported with funds from the Office of Economic Research of the Economic Development Administration, Department of Commerce (Project Number OER-015-G-66-1), and the Program on Technology and Society conducted at Harvard University under a grant from the International Business Machines Corporation. The author also wishes to acknowledge support from the M.I.T.-Harvard Joint Center for Urban Studies. The author takes full responsibility for his views, which are not necessarily those of the sponsoring organizations.

# 7 COMMUNITY HEALTH PLANNING

Marion B. Folsom

Great progress has been made in the fields of medical science and health care during the past forty years, but we have many problems facing us. There is still much to be done in both fields.

The progress of medical science in recent years is indicated by the fact that in the United States life expectancy at birth was 47 years in 1900, 60 in 1930, and 70 in 1964. But it is interesting to note that life expectancy in the Scandinavian countries is above ours. The life expectancy of a male white in 1964 in the United States was 67.7 years; in Denmark, 70.3; in Sweden, 71.3; and in Norway, 71.3. Probably among the factors responsible for their better records is the absence of slums in these countries. We realize the need to make much greater progress in eliminating the slums, raising the income of people with low incomes, and providing better health care for them in the meantime.

The improvement in life expectancy has been due, to a large extent, to a sharp decline in the death rate in the younger groups, with much less improvement in the older ages. There has been a phenomenal decline in the death rate from such communicable diseases as tuberculosis, pneumonia, and influenza. For instance, even as recently as 1920 there were 148,000 cases of diphtheria in this country; in 1964 there were less than 300. There were over 50,000 cases of polio reported in 1952, and only 122 in 1964. There has also been a sharp reduction in infant mortality, although here again our record is not as good as in other countries, particularly the United Kingdom, Switzerland, and Sweden.

In certain chronic and other diseases the trend has been in the reverse direction. The death rates from heart diseases and cancer have increased considerably in recent years.

84

This over-all progress in prolonging the length of life and practically wiping out many diseases has been due in part to the general improvement in the standard of living and to the reduction in the number of families with low incomes. But important factors have been medical research and the improvement and wider use of medical and health services. Now that we have gained so much knowledge from the rapid and extensive expansion of medical research, our concern is to make it available to practitioners and other people so that all can benefit.

Medical research expenditures are now running about ten times the level of ten years ago. In spite of the rapid increase, we are still spending a comparatively small amount on medical research in relation to research in other fields. The greater part of research is being done in the laboratories of medical schools, financed largely by the federal government and carried on by scientists and medical school graduates, often assisted and encouraged by scholarships and training grants from the federal government. Many scientists are now active in the heart, cancer, mental health, and other fields where progress has been slow. As more scientists become available, we shall continue to expand medical research, with greater effort being placed on these chronic diseases, and with the expectation of lower incidence.

We have also greatly expanded our means of delivering health and medical care. There has been a spectacular increase in the construction of modern hospitals, with comprehensive facilities and organization for those severely ill and requiring operations and attention by physicians. Most of the hospitals in the United States (except mental institutions) are voluntary and nonprofit, with construction financed by contributors, with assistance by the federal government, and partly by borrowed funds.

The people have been enabled to use hospitals to a much greater extent, due to the widespread adoption of voluntary hospital and surgical insurance, financed jointly by employers and employees and covering workers and their dependents. About two-thirds of the population in this country now has this protection. Recently, the federal government enacted legislation to cover all persons over 65 for health insurance, through the Social Security system, with the cost to be met on a contributory basis.

Although we have made progress in improving health care, we are still facing many problems in this country in reaching our goal of good quality health and medical care for everyone. We and other countries need far more physicians, nurses, aides, and medical technicians. Steps have been taken to expand teaching facilities here, but it takes several years to train these professional and technical people. In the meantime, we must utilize those already trained more effectively.

Health conditions and health care in the low-income areas of large cities and in many rural areas are much below the level in the rest of the country. Services are generally available in the city areas but the problem is getting these disadvantaged families to make use of the services. Much effort is now being made under the government's antipoverty program to improve this situation.

A most serious problem is the rapidly increasing cost of medical care, especially costs of hospitalization. A good part of the increase in hospital costs has been due to the wages and salaries catching up with the level in other fields, but there are other causes. Many patients are treated in the most expensive acute hospital beds when other facilities, which can be operated at lower cost, would be more appropriate to their needs. Much greater use could also be made of out-patient services, self-care units, nursing home facilities, and organized home care. Most health insurance plans, covering only in-hospital services, tend to encourage the use of these more expensive facilities. The reimbursement of the hospitals by the insurance agencies on a cost basis provides no incentive for the hospitals and the medical staffs to control expenses.

One of the principal factors in the failure to provide good quality health and medical care in many areas and to control costs without affecting the quality of care has been the lack of planning of health facilities and services. During the last few years, much attention has been given in this country to better health planning by the federal and state governments and local communities. Some health problems, such as air and water pollution, affect the whole country or large regions. The federal government has taken steps—through research and grants to states and localities—to assist in abating pollution, but much yet remains to be done.

For many years the federal, state, and local governments have been active in other environmental health fields—such as the protection of the community against impure food, milk, and water, and improper sewage disposal—and in providing public health services. There are in this country literally hundreds of national, state, and local voluntary health agencies concerned with the prevention and cure of specific diseases. They have been effective in providing funds for research and adequate care facilities and in stimulating action by federal and state governments where needed.

In recent years, particularly rapid progress has been made in the treatment of mental diseases. With the use of certain drugs, resulting from research, a decided change has taken place: it has been found that shorter stays are required in institutions and that many persons can be adequately treated on an out-patient basis or at home. There is also a trend toward general hospitals establishing mental health units as a part of the hospital.

The public health officials and the leaders of these national health voluntary agencies realize the government must provide leadership in health and must take a large share of the responsibility and furnish adequate funds in such fields as medical research, teaching, mental health, and pollution. In the final analysis, however, they feel that the health services, both for prevention and cure, must be provided in the local communities where the people live.

In 1962, the federal government and certain voluntary foundations provided funds for a National Commission on Community Health Services, sponsored by the American Public Health Association and the National Health Council, to conduct a study and to make recommendations about how a community could improve its health services. This study was supervised by thirty experienced commissioners, drawn from the U.S. Public Health Service, the medical profession, the health and hospital agencies, and laymen. The Commission was assisted by six task forces of experts in specific fields, such as environmental health, manpower, and community services. Twenty-one cities of various sizes in various sections of the country were selected to conduct a self-study and action program to see what could be accomplished with the guidance of the Commission's staff in improving local health services.

The results of this four-year study were published in 1966 in a report appropriately entitled, "Health Is a Community Affair." When this report was presented to President Johnson in April 1966, the president stated that he could guarantee close and careful study of the report.

I can only summarize here the principal recommendations made by this National Commission as to how a community should plan and coordinate its health facilities and services. The entire report was directed toward the local "community of solution," as the Commission termed the local area where complete and coordinated action can be taken. It suggested the following:

Each individual should have a personal physician, who should see that he obtains the proper preventive services, the appropriate care when ill, and rehabilitative care and services. Each physician should have access to a hospital, preferably a teaching hospital, affiliated with a university medical center.

There should be closer cooperation and coordination of the various specialists and other physicians, through group practice, which would provide better quality of care and better utilization of skilled personnel. Much greater effort should be exerted by the educational institutions—medical schools, dental schools, schools of nursing, community colleges—to increase the supply of health manpower.

There should be better utilization of health manpower by hospitals and other health agencies through improved personnel management, better recruitment, training, and development measures, and more use of part-time workers. The Commission urged closer cooperation among hospitals through joint efforts, such as centralized laundries, laboratories, purchasing, and computers, and reduction of unnecessary duplication of services and little-used equipment—all of which would conserve manpower and lower costs.

Greater use should be made of out-patient services, self-care, rehabilitation, extended care, long-term care units, and organized home care services, to relieve the expensive acute hospital beds and provide the level of care best suited to the needs of the individual patient.

Physicians should cooperate with hospitals in the formation

of effective utilization committees to see that proper use is made of these various levels of care. There is need for greatly improved health care for those living in low-income areas where incidence of certain types of illness is unusually high. There should be a rapid expansion of voluntary health insurance to those not covered and for medical costs not generally covered. Better means are needed to assure pure water, clean air, safe food, hygienic housing, and adequate recreational facilities.

Mental hospitals should be closely integrated into the community, with more patients treated on an out-patient basis; more general hospitals should have psychiatric units, both in-patient and out-patient. Each health agency, voluntary or governmental, local, state, or national, should have a periodic, objective review made by an outside competent group to see that its program and organization are up-to-date and effective in meeting current needs. Each state should have a strong health planning council which could establish and coordinate regional and community planning councils. The federal government should make grants to assist in state and community health planning activities. Each community should establish a strong, broadly representative planning-action council to develop a coordinated community plan for comprehensive health care and to obtain action on measures agreed upon.

Finally, success in achieving the goal of comprehensive health care requires that citizens become well informed and motivated to follow sound health principles and to make use of health services for the prevention and treatment of illness and disability. The Commission's report has had wide circulation and many communities have been stimulated and assisted in organizing local community planning agencies.

As a sequel to the completion of this study, the federal government in 1966 passed legislation providing for funds for the establishment of comprehensive health planning agencies at the state, regional, and community levels. This program is just now getting under way. While a few communities have made good progress in recent years in community health planning, this legislation—with part of the funds being made available by the federal government—will stimulate the rapid establishment of these local planning agencies.

As an example of a community planning effort which has been in effect for several years, I will describe the activities of the Rochester, New York, area. Twenty years ago a foundation provided funds to establish a regional hospital council in the eleven-county area adjacent to Rochester, with a population of over one million, to demonstrate the value of regional planning. Thus the hospitals already were accustomed to cooperating with each other. They had established many cooperative services, such as purchasing, credit and collections, and administration, through discussion groups composed of the staff people of the several hospitals.

In 1963, with the aid of a grant from the U.S. Public Health Service, an active planning committe was set up to conduct long-range planning for the region. Complete data have been collected regarding current and projected population and health needs, individual hospitals and communities, and other pertinent topics. Good progress is being made in developing and maintaining a master plan into which specific projects can be fitted.

In 1961, a council was established in Rochester to allocate funds, which were to be raised in a joint capital campaign, for seven hospitals in Rochester and the surrounding County of Monroe. It was contemplated that the council would also serve as a community health planning council. The group, called the Patient Care Planning Council, was composed of top representatives of the hospitals, the other health agencies, the organization responsible for raising funds for the voluntary welfare and health agencies in the community, the medical society, the University of Rochester Medical School, the county, the city, labor, and employers. Top leaders were selected not only to develop health plans for the community but also to obtain action on the plans developed.

The first step was a scientific survey of the patients occupying about one thousand beds to determine the type of facility or service they needed for medical reasons at that particular time. For the study, the medical society cooperated in obtaining 28 physicians—half from within and half from outside the community—with an internist and a surgeon visiting each patient, consulting the doctor and nurses, and examining the records. The study revealed that a substantial percentage of the patients (23 per cent of the medical patients and 11 per cent of the surgical cases) should not have been in the acute hospital for medical reasons but could have been cared for in the out-patient departments, a long-term care facility, a self-care unit, in the organized home care program, the doctor's office, or simply at home. If optimum facilities were available, an additional 5 per cent could have been transferred.

A thorough survey was also made of the existing hospital facilities and projected community needs. Based upon these surveys, an agreement was reached on a plan which reduced considerably the number of additional acute beds the hospitals had requested. Instead, three hospitals would construct units especially suited for rehabilitation and convalescence, which could be constructed and operated at a lower cost. With fewer beds added, more of the funds were available for modernization. Several old buildings were torn down and replaced by modern wings and equipment. After much discussion, the council was able to obtain unanimous approval by the voluntary hospitals on the proposed allocation of funds.

We avoided the mistake of many communities where each hospital has raised funds independently, without considering adequately the needs of the whole community, which frequently results in an excess of acute beds and a shortage of other facilities, and unnecessary duplication of equipment. The publicity arising from this thorough study of needs contributed to the success of the fund raising campaign, which exceeded the goal by 10 per cent.

Instead of replacing old laundry equipment in three hospitals, a centralized laundry has been constructed for five hospitals, with borrowed funds to be paid off from charges and an expected saving in unit costs. The hospitals have participated in joint purchasing for several years. We are now studying the feasibility of coordination of special laboratory equipment.

The community college inaugurated a two-year nurse training course and conducts preclinic courses for the first year of the hospitals' training programs. As a result, one hospital decided to abandon a projected new nurses' dormitory and its own school program. An organized home care program has been conducted for several years to provide nursing, medical, and other care for patients who no longer need to stay in the expensive acute hospital beds. The average daily cost is about one-fourth the daily hospital charge, and it is estimated that about half of those cared for would have occupied beds in the hospital without the program.

The local medical society encouraged hospitals to organize utilization committees to assist in having ill persons taken care of in facilities or with services best suited to meet their specific needs. A county mental health council was established to coordinate agencies in the mental health field, including the large state hospital. A general hospital has added an in-patient psychiatric unit. Three other agencies combined with the hospital to develop and operate a mental health center, which provides various services.

As a result of studies by the Patient Care Planning Council, the Municipal Hospital, for indigent patients, was incorporated into the University of Rochester Medical Center; the County Tuberculosis Hospital was abandoned and the patients transferred to available beds in the nearby state hospital, with considerable net savings; and the University Medical Center has agreed to provide medical care to patients at the County Infirmary in order to improve the quality.

In 1966, the organization of the voluntary health agencies in the community was combined with the planning council to form the Health Council of Monroe County, as a comprehensive health planning agency. This Council now has subcommittees, composed of professionals in the particular field and interested laymen, assisted by a staff man of the Council, working in the following areas:

1. health manpower, to stimulate the teaching, training, recruitment, and development of health personnel;

2. long-term care, extended and other levels of care to relieve acute beds;

3. improvement of health conditions in the low-income areas;

4. expansion of voluntary health insurance;

5. liaison committee with the medical society to study the possibilities of group practice;

6. environmental health committee to study means of abating air and water pollution;

7. financing committee to assist in financing of capital improvements.

92

The work with the hospitals—voluntary and governmental—and the activity of the Mental Health Council will continue as before.

As these two health councils—the regional and the county—in the past were voluntary and advisory, they could not enforce their recommendations and had to depend upon persuasion to convince the agency concerned. In most instances they were successful, but in a few they were not.

While voluntary community planning councils have accomplished much, without authority a council cannot always be successful in convincing a voluntary agency that the needs of the community should come first.

For these and other reasons, in 1965 New York State passed legislation which gives authority to the regional planning councils to determine whether proposed hospital construction or equipment are needed in the community, and the State Health Commissioner is guided by the recommendation of the council.

With the funds provided by the recently passed federal legislation for planning grants, many regions and communities throughout the country will be establishing health planning councils. We can therefore look forward to much more rapid progress in providing comprehensive services for the prevention and treatment of illness.

# 8 POLLUTION: ANOTHER DIMENSION OF URBANIZATION

**Glenn R. Hilst**

In July 1959 a very large number of people across the United States enjoyed, on television, one of their favorite pastimes, the annual All-Star baseball game. In the midst of this exciting ball game the television announcer suddenly noted that the smog had lifted! The cameras dutifully turned their gaze beyond the ball park, and, sure enough, the mountains a few miles from the ball park were visible, where before there had been only a murky, formless haze. And for a few seconds the national audience shared visually a common experience of the Los Angeles resident: air pollution.

On another scene, residents and visitors to the northeastern United States have enjoyed and prided themselves on the quality and variety of sea food to be found in the rivers and coastal waters of that scenic country. Now, they may find shellfish in their favorite restaurants and supermarkets but they will probably hesitate before ordering these delicacies. Why? Because they are now all too aware that water pollution has invaded the natural habitats of clams and oysters and left them less fit to eat and more prone to carry disease, such as infectious hepatitis.

More recently, Mayor Lindsay of New York City found it necessary to intervene between his Commissioner of Air Pollution and his Commissioner of Sanitation because the rules applied to apartment house incineration of solid wastes—paper, cloth, scraps of food, containers, and the like—by the Air Pollution Department had suddenly resulted in a very large increase in solid garbage which had to be collected and trucked away from the city. The Sanitation Department, which provides this pickup and trucking service, had neither the facilities to handle the increase nor a place to put these wastes. For a few uncomfortable days parts of New York City had unsightly, odorous, and unsafe mounds of garbage on their sidewalks and alleys.

"Pollution" has become a household word in the United States. To the man on the street it is still a rather vague thought, but more and more he is realizing that the air isn't as clean or free of odor as it might be, the water he drinks has strange tastes, and his streams may be a good place to stay away from rather than enjoy. And he is more aware of the all-too-visible garbage dumps, abandoned automobile junk yards, and industrial slag heaps. Perhaps occasionally he wonders, "If I can't burn them, flush them, or just throw them away, what am I going to do with these things I don't want around anymore?" And with this question he has put his finger on the heart of the problem, for it is in identifying things that we "don't want anymore" that we find the genesis of our pollution problems.

In our homes, our businesses, our factories, our transportation systems, everywhere human activities go on, we produce waste materials, things we don't want anymore. Just how much waste we produce depends both upon what we are doing and how frugal we are, but inevitably we produce some things which are no longer directly useful to us; and, because they occupy space, we must somehow get rid of them. These wastes may be the gaseous products of combustion, they may be foodstuffs, they may be scraps of metal, they may even take the form of junk automobiles or abandoned buildings. Depending upon how we decide to throw them away, however, they are all candidates for the pollution of our natural environment, our air, water, and land.

The human activities which produce pollution go on in some form everywhere, and cannot be uniquely identified with our urban areas. However, urban areas are much more susceptible to environmental pollution for two reasons. First, the high-density population and the variety of activities which define an urban area lead to much larger and more varied pollution production. Second, while the production rate is high, the capacity of the atmosphere and the surface waters to assimilate gaseous and liquid wastes is essentially unchanged, and land surface for solid waste disposal is, of course, greatly diminished by other demands for land usage. These are the reasons why environmental pollution has emerged primarily as an urban problem.

However, there is still another fact which is causing the citizen to take action, generally through his local, state, and federal government. Pollution in our cities is beginning to exact a higher price than we want to pay. In order to evaluate this price and what might be done about it, we must examine more closely the nature and magnitude of pollution in our urban areas.

Just how much waste material does a city produce? Needless to say, the answer depends upon several things: How big is the city? What factories are located there? What transportation systems are used? What types of sewage transport are provided? Where is the city located? These and other factors control the production rate of materials to be disposed of. Several examples, however, provide a guide to recognition of the pollution production capacity of a city.

Take, for example, the consumption of food. In one day an average urban citizen of the United States will use about two and one-half pounds of food. In the same day he produces about the same amount of sanitary wastes, and in addition produces another four to five pounds of other solid wastes. On this basis, a city of one million people must be able to process something more than a thousand tons of liquid and solid sanitary wastes and somehow collect and dispose of another two to three thousand tons of other solid wastes each day; this accounts only for the individuals' contributions of sanitary and solid wastes. Manufacturing processes can easily produce several times the waste materials contributed by the daily activities of the citizens of the city.

One way of visualizing this waste production is to note that if the citizens and industries contribute equally, and if the average density of these solid wastes is one pound per cubic foot, and if these wastes are simply accumulated in some convenient location, our typical city would accumulate a garbage heap ten feet high, a half-mile wide and ten miles long in the course of a single year. To say the least, this heap would be an offense to all of our senses and a dangerous garbage pile!

As another example, this time for air pollution, the combustion of sulfur-bearing fuels in New York City now produces about one million tons of sulfur dioxide, which is released into the atmosphere over that city, each year. As a result, the airborne concentrations of sulfur dioxide during poor atmospheric ventilation conditions regularly goes above one part per million. As we shall see, these concentrations are capable of producing serious effects on the health of people, plants, and animals, and are destructive of many materials. Similarly, automobile exhausts are tremendous contributors of undesirable pollutants. Hydrocarbons, carbon monoxide, oxides of nitrogen, and emissions of lead (from fuel additives) are produced at the rate of about one pound per day per automobile. In Los Angeles, this means about 1,500 to 2,000 tons a day, a strength which, when combined with the limited atmospheric dispersion of that area, produces regularly the hazy smogginess mentioned earlier.

These are only examples of the American city's (and citizen's) capacity for producing waste materials capable of polluting the environment. Recent studies of these pollution problems, notably the National Academy of Science report, "Waste Management and Control," and the report of the President's Science Advisory Committee, "Restoring the Quality of Our Environment," have provided much more complete detail, including environmental pollutants not normally associated with urban activities—for example, pesticides and radioactive materials. Both of these studies provide one very significant conclusion: the waste production of our cities is growing faster than the population growth rate. This finding introduces a strong note of urgency in our attempts to control environmental pollution in the urban setting.

But before we discuss what can be done about this problem, we must review the reasons for doing anything, which are the effects of environmental pollution. The production of environmental pollutants has been going on since time immemorial and man has adapted to them without a lot of fuss and bother. Why not now? There are several reasons why we cannot ignore this problem, but one overriding reason deserves mention first. In earlier times man could ignore the consequences of his waste production primarily because he could move on to uninhabited territory practically at will. Clean air, usable water, and unclut-tered landscapes were, figuratively if not literally, just over the horizon. Now we are running out of this frontier land, and just over the horizon is also occupied land. As our population has grown we have occupied the land, and in the process we have become much more conscious of our social as well as our indi-vidual stewardship of the finite resources which the land, air, and water provide for our welfare, safety, and comfort. In the urban setting these features have been magnified many times by the compression of many people and activities into a relatively small area.

This very fundamental point deserves repeated emphasis: the American city is in difficulty with air, land, and water pollution because it has either knowingly or inadvertently overtaxed the ability of these components of the natural environment to assimilate waste materials. One of the reasons for this state of affairs is that the assimilative capacity of the environment is large. The atmosphere can and does readily dilute waste gases to one-millionth their initial concentrations (say, from a chimney) in just a few minutes; a free-flowing stream can dilute liquid wastes, too, in relatively short distances. Similarly, solid materials which are "degradable," that is, decompose in the presence of water, sunlight, and, occasionally, biological actions, can be disposed of in several ways. When underused, these resources can be easily taken for granted. However, in each of these uses of the environment, we must recognize when we are putting in materials faster than nature can reduce them to harmless or inoffensive concentrations, a natural limit which many of our cities have now reached.

The cities that have not reached these limits should take a stern lesson from those that have, and plan their growth and diversification with this limit in mind. For the cities which have already demanded more of nature than she can provide, the first step has to be to get the output of pollutants under control, that is, within these limits. But for these cities, too, future growth must be planned with pollution control in mind.

What happens if we do exceed nature's capacity to assimiliate wastes? What are the consequences of urban pollution? The

consequences of environmental urban pollution may be conveniently grouped under three broad headings: (1) the effects of
pollutants on the physical health of people; (2) their effects on
animate and inanimate objects, effects which have a direct economic impact; and (3) their effects on the activities and attitudes
(or mental well-being) of people. The health effects of environmental pollutants can be and have been extreme. At one time
or another air and water pollutants have killed people. Fortunately, these extreme situations are rare today, but when we
recall the ravages of typhoid fever, cholera, and dysentery, perpetuated by polluted water in our cities, we have a grim
reminder of the potential deadliness of untreated, contaminated
water. Similarly, the air pollution episodes in London, in Donora,
Pennsylvania, and in several other places bear adequate testimony to the ability of air pollutants to exact, as one of their
prices, human life. The most recent, and not yet fully docu-
mented instance of lethal air pollution occurred in New York
City during the Thankgiving weekend of 1966. Approximately
eighty people apparently succumbed to the high levels of sulfur
dioxide and other pollutants which accumulated over the city
during that weekend.

Analyses of the victims of air pollution have pointed up two
facts about these people: first, they were generally either very
young or very old, and, second, the vast majority of the adult
victims were weakened by pre-existing cardiac or respiratory
difficulties. In all of these episodes, normally healthy people sur-
vived and recovered without apparent residual health effects,
although large numbers were incapacitated by respiratory illnesses for several days. These people who were made seriously
ill, frequently to the point of requiring hospitalization for at least
a day or two, illustrate the ability of pollutants to produce illness. At this point, however, we begin to run out of specific
information on the health effects of pollutants by themselves.
Medical records, such as physicians' office calls and hospital admissions, show some correlation with pollution levels, but no
definitive studies which clearly show the role of pollutants are
available. It is generally concluded that only rather high levels
of specific toxicants, such as those encountered in more lethal

99

LEWIS AND CLARK COLLEGE LIBRARY
PORTLAND, OREGON 97219

air pollution episodes, can be clearly identified with clinical illnesses.

Much of the medical and epidemiological research on environmental pollutants in the United States has now turned to attempts to identify potential health effects due to long-term or chronic exposures to lower levels of pollutants both in water and in air. So far these studies have been inconclusive; but in the complex mix of factors which affect human health there is a strong suspicion that prolonged exposure to gases and particulates common to our urban atmospheres has a deleterious effect on health. It does at least seem certain that these pollutants do not improve our health.

Given then that pollutants are adversely affecting the health of our urban citizens, how do we evaluate the price of these effects? On the American scene, health and life are highly valued without regard for the economic or social contribution which any one individual can make. To be sure, there are socially acceptable ways for killing people, but never by any human action against which the individual is defenseless. We kill some 50,000 people each year in automobile accidents, but the individual is free either to avoid riding in automobiles or to drive defensively. Environmental pollution, to the extent it arises from multiple activities over which the individual has no control but to which he is inescapably subjected, is not socially acceptable. And most of the work which has been done to control environmental pollution has been tacitly justified on this basis. In this area of health, economic costs are regularly ignored in favor of human welfare.

Beyond human health, there is a whole catalog of effects of pollutants on things which represent tangible wealth or resources. Plants and animals, too, are susceptible to pollutants and are frequently destroyed or rendered unfit for human use on this account. The difficulties of maintaining ornamental plants in large cities is well known. Less appreciated perhaps are the residual effects of urban pollutants over the surburban and rural areas downwind or drownstream from the city. Inadequate treatment of sewage dumped into urban rivers and streams kills fish, renders the water unfit for consumption downstream, and destroys the recreational value of the stream. Urban air pollutants have caused extensive damage to crops and vegetation beyond the city limits, particularly by the action of photochemical smog.

We also know that air and water pollutants can be highly destructive when brought in contact with structural materials such as steel, stone, rubber, and glass. Similarly, the need for large quantities of clean water in industrial processes frequently requires elaborate purification plants for water taken from polluted streams for this purpose. Another example of pollution effects in this category is the need for excessive cleaning in order to maintain a level of sightliness and sanitation desirable to most urban residents. Particulate materials penetrate buildings and soil everything. They also quickly deface structures, vehicles, and public areas.

All of these nonhealth effects have been studied by economists in an effort to attach a price tag to pollution. The effects are so many and varied that no reliable summary can be quoted. However, we can estimate that on the average a resident of one of our larger cities probably spends, as a minimum, an extra one to two hundred dollars per year as a result of pollution. This estimate does not include any allowance for health effects, and most of these costs are not obvious to the individual. Summed over several million people in any one large city, these costs are very high.

Finally, the most pervasive effect of pollution is not its demonstrable relationship to health or economic well-being, but simply its offensiveness to all the human senses. An unsightly pile of garbage, an offensively aromatic slaughtering house, a grimy window sill or tabletop, the taste of sulfur dioxide, even the noise of uncontrolled manufacturing activities (yes, we have added noise to our list of pollutants), all detract from our concepts of a pleasant and productive environment. For whatever reasons, we are generally uncomfortable with dirt and grime, and we are less energetic and efficient when subjected to these offenses for periods of time. No simple price tag can be attached to these emotional effects of urban pollution, but they are part of a vicious cycle which leads to the final deterioration of a city, the abandonment of unsatisfactory environments by the people who can afford to convert the city to a pleasant and healthful setting and can insist upon these surroundings, leaving only those who cannot afford the luxury of clean, well-kept surroundings.

At this point the character of urban pollution must be identified as the social problem which it is. In the urban setting the cause and the correction of pollution are both primarily the responsibility of groups rather than of any individual. As a determinant in producing acceptable or unacceptable environmental conditions within the city, pollution is one of several factors which must be considered. The economic health of the community, the civic and regional pride of the citizens, the commitment of the community to various forms of industry and transportation, and the willingness of the taxpaying citizens to pay for constant renewal and maintenance of an equitable environment, all these factors finally determine what will be done about keeping the city and its environment fresh and healthful.

These examples of the magnitude and varied costs of environmental pollution as a dimension of urbanization are necessarily sketchy. I hope they are sufficient to point up the need to consider this facet of urban growth and form, both for action now and for planning for the future. The American city has many problems to face, and pollution is definitely one of them. Americans are waking up to the fact that waste disposal is a serious problem, and they are beginning to marshal significant programs to contend with it. By all counts, the required effort is just beginning, but the concern of everyone from the president on down is evident. The relative newness of our recognition of the problems of pollution is demonstrated by the fact that no organized federal program in air pollution existed prior to 1955. Studies of water pollution at the national level date from 1948, and a national program for solid waste studies is just now in an embryonic stage within the Department of Health, Education, and Welfare.

An extensive review of the nature of these problems, where we stand now, and a plan for future action was the assignment made by Secretary Gardner of the Department of Health, Education, and Welfare to his task force on environmental health and related problems in 1966. In its report entitled "A Strategy for a Livable Environment," this task force summarized its findings as follows:

American affluence today contaminates the nation's air, water, and land faster than nature and man's present efforts can cleanse them.

But of even greater concern, experience has shown that undetected environmental health hazards, either alone or in conjunction with known hazards, can arise suddenly to create conditions of living harmful, if not dangerous, to the public. It is necessary then that a constant effort be made to detect these hazards before they reach the crisis stage. But, while we must be alert to the effects of new hazards, continuing effort also must be expanded to learn more about known hazards so that they can be controlled.

We know something of air pollution, but we know little about the hazard potential of 500,000 to 600,000 synthetic chemicals and other compounds on the market today. We know something of water quality, but little of the effects of trace metals. Can we cope with solid waste? What is the future problem of nuclear waste?

The task force concludes that danger to environmental quality, particularly in the broad context that the Task Force has reviewed it, is among the most important domestic problems today. It affects all Americans where they live, work, and play. It can very materially damage their children and generations yet unborn.

What is needed now is an overview of the entire question of environmental health and its interrelated components, not only water pollution, air pollution, solid wastes, but, also, noise, crowding, radiation, traffic safety, and ailments which can be related to these factors.

The American public must develop this overview, a sensitivity to the scope and limitations of the environment.

The public must learn that air, water, and land are limited, that the number of Americans is growing, that their affluence and effluents are increasing, forming something of an ecological chain reaction.

The 140 million Americans who live concentrated on 10 per cent of the nation's land area must now take vigorous action to clean up the environment.

As the facts become clear, the public will be shocked at the price it is paying for its affluence. But, if it is obvious that one way to halt the contamination of the environment is to prohibit automobiles, stop the generation of electricity, and shut down industry, it is just as obvious that this way is impossible. What is possible is to find ways to eliminate contamination at its source. Or, next best, to capture a pollutant and use it in a nonharmful way; or, finally, to bring the level of pollution down to a point compatible with the requirements of human health and welfare.

Three major steps toward the maintenance of acceptable pollution levels were identified in this report:

1. An evaluation of the effects of pollutants and the social and economic costs for their control within acceptable limits.

2. Education of the American public regarding the true costs of pollution and feasible methods for controlling these costs.

3. The development and implementation of the technical and legal means for controlling pollutants at their sources, before they enter the natural environment.

We have already reviewed some of the known effects and costs of pollutants, but a more exhaustive study is required before the complicated interactions of pollutants and human welfare are truly understood. We are now in the position in several cities where immediate action is required to halt the deterioration of air, land, and water. However, today's problems seem almost trivial when we look ahead and visualize the future consequences of insufficient research and planning now.

The urban and regional planners have talked for some time of the growth of gigantic, continuous cities, literally hundreds of miles in length. These cities are to be formed by the eventual merging of our present major metropolitan areas, and all present trends of population shifts to urban areas as well as population growth point in this direction. From some studies of air pollution accumulations over continuous cities I recently calculated that, to a first approximation, the limits on air quality may very well set the limits of the dimensions of such megalopolis and these limits may be no more than a few hundred miles. These calculations are in need of further refinement and experimental verification, but they point clearly to more serious problems in the future. Similarly, water supply and liquid and solid waste disposal loom as much larger problems in the future. Just how serious can be guessed at now, but we need a much increased effort to gather the knowledge and understanding required for more definitive answers to this question.

The program of education of the public about the effects and costs of environmental pollutants is going forward today. And as new understanding is developed, it is quickly translated by the popular news media. This activity is particularly important on the American scene because, ultimately, the citizens decide how much they are willing to pay for the quality of their environment. And regardless of how it is done, it is the citizens who pay, whether through increased taxes or through increased prices on the products they buy. The citizen rightly looks to his government, university, and industrial leaders to inform him of the consequences of pollution, and the ways in which these consequences can be alleviated and at what costs. Then he must decide, through personal action and the ballot box.

As noted first in this report, one of the major obstacles which we must overcome before real progress can be made is to establish reasonable limits on environmental quality. As is true of any situation which affects large groups of people, there is a fairly wide range of opinion as to what is a satisfactory level of pollution. Some would insist upon no pollution at all, and at the other extreme some would not care how filthy the world is around them. The attainable and desirable objectives of environmental quality are somewhere between these extremes. We are trying to find out where the cutoff points should be, both in terms of individual desires or needs and in terms of what we can afford. There is no way around it, pollution control costs money, and expenditures for this purpose must be weighed against the benefits to be achieved.

Given the objectives for pollution control, we have several ways they be achieved. One is simply to insist that pollutants not be released beyond predetermined permissible levels into the air, land, or water. Essentially this would require recycling or permanent storage of most of the materials we now throw away. Since the environment around us can dispose of some wastes harmlessly, we do not have to think in terms of 100 per cent control of pollution sources. The method of pollution abatement at the source also requires the development of devices and technology for capturing pollutants of many different kinds. In many instances this technology exists. However, there are also some pollutants which we do not know how to capture economically and in large quantities. An excellent example is sulfur dioxide, for which we have only expensive abatement techniques. In fact, the National Center for Air Pollution Control, a part of the U.S. Public Health Service, has just launched a massive research effort to find practical ways to control this ubiquitous and insidious air pollutant.

At the present time, control of pollution by retention or treatment at the source is the primary approach being used. Legal devices for requiring such abatement are being developed at all levels of government, including federal, state, and local laws of varying degrees of severity.

An alternative, which has not been developed seriously, is the deliberate design of the form, size, and location of the city so as to minimize pollution problems. As noted previously, there is strong reason to believe that continued growth of our large cities by expansion into the surrounding countryside, until large cities merge into megalopolis, may create cumulative pollution conditions which will cripple the city economically. This possibility needs to be studied, but in the meantime the urban and regional planner must include environmental constraints on his list.

For pollution is, quite simply, another dimension of urban design. It arises from human activities and is either aggravated or alleviated by the types, level, and density of these activities which we build into our cities. These arrangements of human activities are indeed grist for the planner's mill, a challenge to his knowledge and ingenuity, and ultimately a part of his professional responsibility to the countless people he serves.

# 9 URBAN DESIGN

## Wolf Von Eckardt

Urban design is an old art which, I believe, has new and vital social importance today. Unfortunately, not everyone shares this view. Most politicians, builders, and even city planners consider urban design merely a matter of aesthetics that costs a lot of money and starts a lot of silly arguments—a nice but unessential frill. But what is essential?

We seem at long last agreed that we must have a modicum of order in our man-made environment if it is to function properly, if, rather than getting killed by chemical pollution, chaotic automobile traffic, and urban riots, people are to live in relative safety and comfort. That means, we are also agreed, that we must arrange things so as to accommodate reasonably the many conflicting economic and social needs and interests of urban society. But from this point on there is still considerable confusion.

There are, to begin with, bound to be different views of the relative validity of the conflicting economic and social needs we want to accommodate. No mortal is granted divine objectivity. We might therefore just as well acknowledge that, like it or not, we must approach the task of arranging things in the city from a given point of view, a Weltanschauung, a—let's face it—political and ideological platform.

But in this country we don't make that acknowledgment. City planners maintain the notion that arranging the lives of people can be done with the same aloof objectivity as arranging molecules in a test tube. The politicians, therefore, assume that they have nothing to gain from the exercise—and ignore it. This, of course, makes city planning rather ineffective.

Actually, effective city planning and building has always been an ideological and political act (*polis*, remember, means "city"). It has always been a means of either advancing an ideological and political concept or of attempting ideological and political reform. The Greeks built their cities as a setting for the Panathenaic processions; the Romans to advance the ideals of their military-industrial complex; the Middle Ages to symbolize spiritual and civic unity; Pope Sixtus V and his Renaissance planners for the greater glory of ecclesiastic power; Baron Haussmann for a rather more secular version thereof; and the utopians and socialists of the early industrial age to reform capitalist outrages. The latter's ideas, which led them to such utopian settlements as Familistiere in France, about 1860, to Le Corbusier's luckily unbuilt Ville Radieuse, and Ebenezer Howard's luckily built Garden Cities, still dominate our planning concepts and legislation today, although, all too often, conservatives and capitalists promptly, conveniently, and all too superficially took them over for their own ends.

In this country, however, city planners, as I said, still castrate themselves by purporting to stay aloof from ideology and politics. And conversely (with the exception of Franklin Roosevelt's New Deal, which launched public housing and made a half-hearted attempt to build three New Towns), America's political leaders and even ideo-political movements such as labor and civil rights, stay aloof from city planning, leaving the not unimportant question of where and how their followers are to live largely to happenstance and that tiny section of free enterprise that stands to make profits in building, real estate, and mortgage banking. Without the vital fuel of idealogy and the machinery of practical politics, city planning therefore sputters along rather haltingly and ineffectively. But that is only part of the confusion. Another is that we have not quite settled on just how we are to go about arranging things more efficiently in the city even if we had a point of view.

Logic would seem to dictate that we first attempt to find out just what people's needs and interests are. That is urban research. Secondly, we would figure out how to accommodate reasonably these needs and interests in a given space and time.

That is urban planning. And thirdly, we would seek to translate these arrangements into physical structures so that they have the desired effect. That is urban design. The three are obviously closely related and would modify each other in a creative give-and-take up and down the line. Unfortunately this simple and logical sequence more often than not eludes what American city planners so fondly call "the planning process." Here in Washington, for instance, we have just spent more than two years and I don't know how much money on the design of a new highway bridge before it occurred to the planners that research into whether or not the bridge was actually needed had never been done. They have now called for it, and the laboriously designed bridge may never be built.

But what happens more often is that planners confuse research with planning. They would count the noses of the people who went by automobile from point A to point B ten years ago; count noses, or rather car bumpers, again today, run these statistics through a computer which projects how many people might take the trip in another ten years, call for a new freeway to make the trip easier for these projected motorists, and call the exercise city planning. It is not, of course. Computerized crystal gazing is not planning. Whether or not that freeway is desirable never enters the planners' minds, since, lacking any ideological concepts, they have no desires. But they make a mere statistical projection an imperative. The future becomes a mathematical multiplication of the past. And past mistakes are turned into disasters.

This much we are now beginning to see, perhaps, as we discover that a freeway (or a shopping center or housing project or whatever) built in one place has countless ramifications on others. We have discovered that there is an urban ecology, and we may even see that it is not only development as such, but the timing and tempo of development that affects it. This begins to make us more cautious about confusing research with planning. But we still confuse planning with design, which is just as bad but a little more excusable. It is just as bad because mere planning for a workable accommodation does not make the accommodation workable. We have, for instance, planned and built in our cities hundreds of public housing projects that were noble, decent, safe, and sanitary in their planning concepts but so poorly designed that in some instances people actually refused to live in them. They have also demonstrably set back the whole idea of public housing where it is most needed. The plan, alas, merely charts the necessary and desirable human activities in urban space and the social and economic relationships on which these activities depend. It does not automatically bring with it a satisfactory utilization of the space. This is the function of urban design. And it seems to me a rather essential function indeed.

The confusion of urban planning with urban design is, however, understandable (and thus excusable) because in the past the two were essentially the same. If a king or high priest or a social reformer wanted to impose his ideas of how people ought to live and how they ought to feel about his ideas, he instinctively reached for a pencil and a map of the place. And as he doodled, the abstract concept instantaneously translated itself into a visual image, an image determined by the geography of the place, the available technology, and the prevailing style and convention of his time. His planner refined this rough concept. He was the draftsman of the idea, the draftsman who brought about the detailed synthesis of ideology, geography, and Zeitgeist and turned it into an urban design. There was sometimes argument about money. There could hardly be much argument about geography and Zeitgeist. And the king, priest, or reformer was smart enough to make sure that everyone agreed with his basic ideas.

In other words, you had a river and a hill where you wanted a lot of people to live and do business, and you had only circumscribed technical ways to build them a bridge and roads and a protective wall. Popular consensus, or the Zeitgeist, prescribed rather definite modes and styles of building the houses. Everyone, furthermore, was pretty much agreed about what a church and a city hall ought to look like. The location of the marketplace was determined by the distance the housewives in town were willing to carry their shopping baskets. And the king or priest insisted on placing his castle or church up on that hill. So all the planner had to do was manipulate all these given factors to make sure that the utilization of the space was functional and attractive, and that the whole thing had the ideological impact the king, priest, or reformer desired. He was really not a planner in our sense but an urban designer who may have given the city planners of today a better reputation than they deserve. For when it was all built, everyone thought it beautiful, felt strongly attached to the place, identified with it because it was, of course, an expression of the collective identity, and lived happily ever after—until, in the twentieth century, the town got far too big and was invaded by far too many automobiles.

This expansion destroyed the sense of identity and security in the city. Identity and security: these two words, perhaps also expressed in the word "shelter," are, I believe, what people seek in a home and the extension of the home, the town. For the town, rather than "the housing unit," where we sleep and sometimes cook and watch television, is the place where we live, work, learn, worship, relax our minds and bodies, and try to make a go of civilization. That is why I feel it to be a mistake when we put all our emphasis on housing, to the detriment of the planning and building of the town, the community.

In the old town of our hypothetical king, priest, or reformer the sense of identity and the sense of security were, so to speak, built in. They were built in because the town was the visual expression of the order of society. You could see where God lived and where the king or the mayor lived and you could experience community in the *agora* or marketplace and solitude in the temple, church, or town park. It was the sight of it, the sight of an orderly and comprehensible relationship of accepted values that made you realize who you were as you looked at that spire, that hill, and that river. And it made you feel secure because you were part of it, part of a whole. The orderly, comprehensible, and stimulating design of the town gave it what today we call "a human scale."

One need only leaf through a modern city planning report, probably a rather voluminous document, to realize that the theoretical accommodation of urban needs and interests—the prescription for housing densities, traffic patterns, open spaces, and all the rest—even if it is absolutely brilliant, is far removed from this visual expression that is apt to command our loyalty and will cause us to live happily ever after it is all built. This is still true if the planner, for the benefit of the politicians who don't care to read long documents, presents his plan in the form of a three-dimensional, schematic model. The planner is no longer a designer. He is much too busy worrying about densities and facilities, traffic patterns and open spaces to worry about the intricate forms and shapes, light and shade, texture and color, vistas and enclosures, let alone such all-important amenities as pavements and benches and light fixtures that bring the plan to life and evoke the desired human response.

And this, at last, brings me around to answering my own, initial question as to what, in addition to functional efficiency, is essential about building and rebuilding our urban habitat. It is, I believe, the all-important human response to that efficiency. If the arrangement is to be successful, people must like it. They will never like the cubic feet in a housing project or the per hour capacity of a freeway. But they may, largely for unconscious reasons, respond very favorably to the ambience of a new neighborhood and the way the looks of that freeway pleases them. We take an undue risk, therefore, if we stop with the effort of urban research and urban planning and leave urban design to chance. And yet this is what we usually do in this country.

In our government-sponsored urban renewal efforts we stop with a land-use plan and a schematic model and then turn the job of constructing the desired structures over to private developers and their architects. Their contracts bind them to follow the plan, of course. But they are not required to correlate its bits and pieces into a coherent whole. They could not care less about urban design. The developer is interested in making his building pay. And the architect is interested in making his building great. The one is in business for profit, the other for fame. Both are laudable aims. But an agglomeration of profitable monuments to private builders and architects does not assure the public interest in the viability of the effort. The undesigned agglomeration rarely adds up to a socially responsible and response-evoking place for the rest of us.

Eliel Saarinen put this very succinctly: "Were a great number of the most beautiful and famous buildings in architectural history all re-erected to form a single street," he has written, "this street should be the most beautiful in the world, were beauty merely a matter of beautiful buildings. But such would most certainly not be the case, for the street would appear a heterogeneous medley of disrelated edifices. The effect would be similar to that produced if a number of the most eminent musicians all played the finest music at the same time—but each a different key and melody. There would be no music, but much noise."

This judgment is particularly true at a time when builders feel little restraint from social purpose and architects tend to seek glory in originality. And the visual noise is particularly bothersome at a time when a technology and an affluence that make it even possible for us to make war on the stars are able to overcome the traditional restraints of geography and architecture. We can bulldoze the hills, fill in the rivers, put everybody on wheels, and build mile-high megastructures. And don't think we are not sorely tempted. Perhaps this Promethean pride, the creed of the technocrat, is our new ideology. But it is not, I believe, the kind of ideology that will build good cities.

Evidence for this belief is the fact that the technocrats, and certain city planners who have been seduced by them, would abolish the city altogether. Buckminster Fuller, for instance, rejoices in the possibility that we can all divorce ourselves from the earth, and the cumbersome sewers and water mains we must now dig into it, with the help of a little black box such as astronauts use. The box regenerates our wastes and water and even reconditions our air and provides us with light and heat. If only we strap those little black boxes to our backs, he says, we can all disperse over the world's mountains and deserts, telecommunicate with each other, and dispense with crowded settlements. Fuller, needless to say, did not acquire his astounding, sophisticated knowledge from video screens on lonely mountaintops. He acquired it in the lively bustle, the intellectual interchange, and the accumulation of wisdom that crowded human settlements stand for.

And that is what obviously keeps attracting others as well. The

Nazis all but completely destroyed Warsaw. This city of one-and-a-half million people was so utterly devastated that, when the shooting was over, the planners debated whether there was any sense in rebuilding it. After two or three weeks of deliberation they looked out the window and saw that a quarter of a million people had already returned and started to dig the city out from under the debris with their bare hands.

Every day of the week planners in both the developed and developing countries, in countries where the population is exploding and in countries where the population remains static, look out of the window and see more and more people flocking into the city. The only way we can keep them out in the country, it would seem, is to build them New Towns out there that meet their longing for an urban environment. Deserts and mountaintops are nice places to visit. But one of the troubles of our time is that everyone wants to live in or near the city.

But how do we design good cities? I find it fascinating that, in search of the answer to this question, modern urban designers inevitably turn to the past. Inevitably they try to understand what made old cities so attractive, so civilized, so human, so livable. Camillo Sitte, in fact, came right out and said it. We have long since lost any natural sensitivity and instinct for building a good place to live and must thus rediscover past principles of urban design. Sitte's search for these principles soon led him to the ideological motivation behind these principles which I mentioned before. Design is a process of decision-making. The decision, for instance, to place a church within a row of buildings rather than in isolation is not only aesthetic, functional, or economic. Aesthetically you might want to keep a plaza open or help enclose it. Functionally you may want direct access to adjacent buildings or not. Economically you may want to save the cost of building an attractive facade all around. But it is also an ideological decision. Is the church to be set apart from mundane life or is it to become part of it?

This is not the time to enumerate these past urban design principles and their ideo-political starting points. This is not a historical lecture. But it is interesting to note that, in accordance with Kant's dictum that the nature of the object is determined, at least in part, by the nature of the subject look-ing at that object, accounts of past urban design principles, too, are largely determined by the subjective view of the his-torian who looks for them. Relativity and the good old Zeitgeist again! The most significant recent account is that by Edmund N. Bacon, the director of the Philadelphia City Planning Com-mission. In his brilliant book, *The Design of Cities,* Bacon recognizes and stresses that the important thing about urban design is not how the city looks all at once, but how it affects us as we move through it. His organizing concept is thus the movement system, with all its relevance to our mobile time. But "the changing visual picture," says Bacon, "is only the beginning of our sensory experience; the changes from light to shade, from hot to cold, from noise to silence, the flow of smells associated with spaces, and the tactile quality of the sur-face underfoot, all are important in the cumulative effect." And all, I might add, can only be the result of creative urban design—the vital link between city planning and city archi-tecture.

Obviously, then, urban design circumscribes the work of the individual architect. But it will never stifle his creativity pro-vided we don't confuse creativity with originality. Quite to the contrary. The architect will be more creative, not less, if he can work within the discipline of an over-all scheme. The com-poser's score has never yet subdued the creative brilliance of a talented musician, though in interpreting Beethoven some pi-anists will at times also somewhat modify his composition. By the same token good urban design may also at times modify the city plan. Design may make possible things the planner never thought possible. In one American city, for instance, they thought that surely nobody would want to live along a certain street opposite an industrial plant. As it turned out, the plant, surrounded by handsomely landscaped green space, proved a special attraction to people because it was so well designed. Things often depend not on what is done, but how it is done. This planners often forget.

In the orderly, logical planning process I advocate, the plan-ners would therefore stick to their task and not predetermine design decisions. If they prescribe high density of residences in a given area, for instance, they should do just that and not prejudice the design by insisting on a conventional pattern of high-rise buildings. The designer may be talented enough to find other ways of accommodating these densities—with terraced housing, for instance, or a mixture of high and low buildings, or an inspired structure like Moshe Safdie's *Habitat* in Montreal's Expo '67. Creative design, I believe, is given far too little opportunity to get us out of the present urban rut.

And I also believe, although no political leader in this country has yet seriously written them on his banner, that we have in recent years begun to evolve the essential principles for a creative urban design for our day. One, of course, is the principle of separation of different modes and destinations of movement. We must liberate the poor pedestrian from the hazard of automobiles, just as we have long agreed to protect him from railroads and vice versa. And we must separate fast-moving through traffic from slow-moving local traffic. While the transportation planners are beginning to understand this, they have not yet learned to work closely with urban designers on ways to integrate their transportation network with the urban fabric of buildings and open spaces. It is easy to keep traffic moving in and out of the city on its own roads. The more difficult problem, which we have not fully faced yet, is how, as the English traffic expert Colin Buchanan put it, cars and other conveyances are to efficiently "wriggle" in and out of buildings.

It is part of the problem of separating mechanized traffic to integrate human activities. Much of our trouble is that in most American cities and suburbs we have to mobilize the power of 250 horses to get a pack of cigarettes or a glass of beer. Essential human activities have to be brought into human reach, the reach of our own two feet. It is perfectly silly that with few exceptions, such as Le Corbusier's *Habitation,* we fail to utilize the perfect modern device for this—the high-rise building. It allows us to put stores for cigarettes and such, or beer parlors on the ground floor; offices and perhaps even schools and other services on the next half-dozen floors, and residences above them.

116

Another is the principle of scale. We are beginning to agree that the city has become too vast, too anonymous, too dissipated. We must return to a comprehensible, manageable, and livable human scale. This means new towns of deliberately limited size in the country. And this means new-towns-in-towns within the existing city, nucleating the city as it were, breaking it up into its organic communities and neighborhoods. It also means ensembles of smaller structures rather than the monolithic trade, cultural, government, and corporate centers that bust our cities with their arrogant giganticism.

But the important and related principle for both, the new town out in the country and the new-town-in-town, should be a new and strong emphasis on equalizing the attractions of the city and its benefits among all its residents. Congress, back in 1949, recognized the right of every American family to a decent home in a suitable living environment. A suitable living environment, as I see it, includes nearby shopping, playgrounds, parks, amusements, and cultural facilities, as well as schools, firehouses, and police stations. It will be a long time, I fear, until we can abolish the hardships of poverty by providing everyone with an adequate income. But we can certainly and far more easily give everyone equal access to the public facilities for recreation, education, culture, and beauty that an affluent society can provide.

In the big city this would also include equal access to the department stores, theaters, opera houses, concert halls, museums, sports arenas, and other things which only the big city can offer. We cannot, to be sure, move all people close to these attractions. But we can move these attractions close to where less-affluent people live. I deplore concentrated cultural centers, which we always manage to place within easy access of the rich. The whole city ought to be a cultural center. And I agree with the German architect Hans Scharoun and others who would weave cultural institutions, large offices, and stores as well as parks, like ribbons through the city to make them more equally accessible to everyone. These are, of course, planning concepts. But the principles must extend into an urban design which would make them work, articulate and dramatize them much in the medieval town planning tradition.

We seem to respond to this tradition, again, not because of any false romanticism, but because it has mastered the art of harmonizing a variety of shapes within the framework of an over-all discipline and because of its humanism.

If we will master our technology in the service of such humanism—and this seems to me the insistently growing will of people who think about the environment—the chances that the present crisis of the city will be the beginning of a new age of beautiful cities seem favorable indeed. And if Churchill is right that we first shape our buildings and then our buildings shape us, great cities may then indeed be the beginning of a great society. The intellectual conditions seem increasingly right. We have been called an "affluent society." But that is actually a misnomer. A great many individuals in America today are more affluent than ever before. They live in urban areas. And they demand a better environment—health, convenience, comfort, contact with nature, and spaces and buildings where man can graciously occupy his growing leisure with continuing education and cultural pursuits. They now demand not only individual riches but public affluence, so that we may indeed have an "affluent society." These new urban demands must be urbanely designed for if we are to give, as architect Frei Otto defined it, our cities "a stimulating order." We have, I think, the knowledge and skill, and we surely have the resources. All we need is the daring.

# 10 A SOCIOLOGIST LOOKS AT HOUSING

**Robert Gutman**

The sociologist is concerned with improving the quality of societal functioning. He tends to assume that every element of culture somehow contributes to the excellence or the poverty of the total society, and he includes housing among these elements. Therefore, he naturally wonders whether changes in housing might not have effects which would redound to the benefit of entire cultures, a hope which I am sure is shared by everyone. The sociologist, in other words, when he examines the effects of housing and related aspects of our society, sees himself as a kind of cultural doctor who will be able to prescribe the appropriate therapy for the sick societal patient. In this case, the therapy we have in mind is some alteration and improvement in that element of culture called housing.

The culture of any society is usually thought to be of two kinds: material and nonmaterial. Material culture includes the man-made phenomena which have physical properties such as height, breadth, and weight, which are visible to the eye, and which can be touched. A boat, a machine, a house, a factory —all these objects are regarded as part of the material culture. The nonmaterial culture is that portion of the environment which surrounds man and which has an impact on his behavior but which lacks these material properties: values, beliefs, norms, traditions, and all the other habits and ideas invented and acquired by man as a member of society.

Contemporary sociological theory tends to assign primacy to the nonmaterial culture in choosing problems for study. It assumes, for example, that boats, planes, automobiles, and so forth, are not nearly so important as the traditions we have developed which make their manufacture possible—indeed, which prescribe how we are to use them. The emphasis of contemporary sociology is to insist that the material culture would not exist had not the nonmaterial culture first been available to suggest the ideas which are embodied in the inventions of material culture. This emphasis of contemporary sociological theory is not a matter of professionl ideological perversity. Rather, it seems to be true that a wide range of nonmaterial culture traits is compatible with the existence of any single object or form within material culture; or to put it in another way, material objects do not exert a very compelling force on the development of nonmaterial culture. It is important to keep this general characteristic of the relationship between material and nonmaterial culture in mind as we examine housing.

I have mentioned the distinction between material and nonmaterial culture in order to suggest to you, first, that in the perspective of sociology housing belongs to the material culture; and, second, to emphasize the need, when we consider the effects of housing, to distinguish very carefully between the influences which are the consequence of the material culture and those which are the result of the norms, values, and habits associated with the use of housing. For in the case of housing, as in the case of the wheel, the machine, and the automobile, the physical object never exists without the nonmaterial culture which prescribes how the object should be used.

Is housing really an element of the material culture? I think it is, providing we restrict the definition of "housing" to the physical dwelling per se, with a certain cubic space, with a particular form, with a certain floor plan, and made of materials having a particular texture. To define housing in any other way, to have it include more than the house itself, makes it difficult to locate the boundaries of the phenomenon whose social effects we want to know, either as social scientists interested in understanding the impact of housing or as policymakers advocating the kind of housing that will better meet the needs of the public.

Unfortunately, both in social research and in the professional circles concerned with housing policy, the distinction between housing as a physical object and ideas and prescriptions about the use of housing often is ignored. There are many examples of the confusion which the failure to make the distinction helps to sustain. One outstanding example, about which a considerable critical literature now has developed, is the debate over the effects of American suburban housing on sociability patterns in the United States. The intense neighborly social interaction that is said to occur among the residents of housing developments was for a long time attributed to the design of the suburban house and the site planning which governed the location of the houses in relation to each other. A number of sociologists have pointed out several facts about this pattern of sociability which lead one to question the degree to which it is a consequence of "housing." For one thing, it appears that the families who choose to move into the housing developments, rather than to remain behind in the cities or to move to established suburbs, seek the very sociability opportunities which the suburban house was said to induce in them. Secondly, in his studies of Park Forest, Illinois, and more recently of Levittown, New Jersey, Herbert Gans has been able to show that active, friendly sociability emerged as a pattern in these communities only when the residents of adjoining dwellings shared common tastes and values, were homogeneous with respect to race and class, and followed similar child rearing practices.[1] Gans's conclusions have been confirmed by the research of other sociologists: for example, Bennett Berger, in a study of a California suburb, showed that where a working-class population is not inclined by background and experience to engage in easy sociability, suburban housing does not lead to the kind of behavior usually associated with development living.[2]

Additional illustrations will occur to you, some taken from the literature on public housing in the central cities of America and Europe, some from the literature about new town housing in England and some, surely, deriving from your own experience living in houses. What in general are the implications of such illustrations? What in general can we say about the impact of housing, part of the material culture, on human social behavior, part of the nonmaterial culture? Here are two broad generalizations, and some examples to illustrate and clarify their meaning, which summarize the burden of social science evidence on this matter at the present time.

First, the impact of housing varies with the kind of social behavior that is the subject of an impact. I mentioned earlier that a wide range of nonmaterial cultural traits is compatible with any single object or group of objects making up material culture. It is also true, however, that given any particular object, and any particular nonmaterial cultural context surrounding it, some elements in the nonmaterial culture will be more responsive to the properties of the object, or to changes in its properties, than are other elements. We believe, for example, that the automobile has had considerable influence on the style and location of American courtship practice—on which its effect apparently was immediate and direct beginning in the 1920's —whereas its influence, say, on our method of electing congressmen has been virtually nil.[3]

The principle which operates with respect to the automobile is applicable in a general way to the dwelling unit or house. Houses of various sizes, shapes, textures, and design exist within a single society, while houses similar in physical contours and characteristics are found in different nonmaterial cultural settings. However, although there may be few nonmaterial culture traits which are associated exclusively with particular kinds of housing, housing and changes in the properties of a house do appear to have some influence on behavior. To take an example from the literature dealing with low-income housing, there is some evidence that the provision of more adequate space in the dwellings of the poor has had a salutary impact on the study habits of lower-class children who are attending school. It also should be noted, however, that improved housing has not had an equivalent impact on the occupational success or position of the adults who live in new dwelling units, nor on the delinquent behavior of adolescents.[4] Once the fact is mentioned, it is fairly obvious that housing will have an impact on some kinds of behavior and not on others. How do we distinguish between these classes of behavior? Which behavior will housing affect, which social action is it unlikely to influence? The answer seems to be that housing will influence social behavior only when two conditions have been met:

The behavior must take place within the house. With certain exceptions that will be discussed later, the house, like any other object in the physical environment, has very little effect on behavior except when that behavior is carried out in the presence of the object. This notion is simple, but it also is useful because it helps to explain why a child's study habits can be influenced by the dwelling unit more readily than an individual's place in the occupational structure or his performance on the job. A good deal of studying, after all, takes place within the domicile; work and the job are located in another physical structure, the office or factory.

The behavior must be "housing-specific." Housing-specific behavior is social action which requires the properties of the house as a facility for its successful completion. Lighting, for example, not only is useful for reading, but also is essential for the act of reading to be carried out successfully. Increases in lighting, better air circulation, and more space therefore can have a positive influence on studying behavior. The last sentence is important: when we refer to social action as housing-specific we mean that the facility represented by the dwelling unit is a necessary, but hardly a sufficient, condition for the action to be completed successfully. After all, a child might not be motivated to study and consequently an improvement in the level of illumination of the house will fail to affect his behavior. (And perhaps we ought not to forget that if a child really wants to study, he might be able to do so efficiently by going to a neighborhood library, regardless of the level of illumination at home. In this case we are dealing with behavior which is housing-specific but still is not influenced by housing for the reason that it does not meet the first condition we have named: it does not take place within the house.)

I have no ready list to offer at this time which catalogs behavior according to whether or not it is housing-specific. However, it does seem possible to think of forms of behavior which are more independent of the properties of housing than is the act of studying. Consider crimes like rape or assault, or even property damage, which occur in dwelling units and which housing reformers during the New Deal period hoped could be diminished or eliminated by improving the properties of the dwelling. The passion which leads to rape is a fierce, intense one and there is very little latitude in the construction of the dwelling unit that will protect a victim from an insistent invader, except perhaps to install prison locks on every bedroom. Damage to property is easy or difficult regardless of whether the dwelling is spacious or constricted. Assault, like rape, is not an act against the dwelling unit itself but against its inhabitants, and therefore there are serious limitations on the role which the house as a facility for action can play in controlling or encouraging assault, except, again, if it is constructed to prevent any intrusion.

In my previous remarks I have emphasized the view that the house is best regarded as an element in the material culture. One thing that we can say about the objects which comprise the material culture, regardless of the nonmaterial culture surrounding them, is that they are not simply strewn about, ready to be picked up and used by whoever may come upon them. In every society, the nonmaterial culture prescribes which individuals have rights to use the object, under what conditions, and for how long a period.

Housing, in other words, is a possession, and like all objects therefore contributes to our evaluation of the social class position of the individual or family who possesses it. Especially in complex modern societies, there is a pronounced tendency to measure a person's place in the class system in terms of material possessions. Housing is one of these objects which is taken into account in our judgments of others. Not only do we judge others in terms of their possessions, but we tend to evaluate our own social position reflexively according to the objects we possess. Therefore housing represents one criterion for allocating individuals and families to social roles; and housing also contributes to the image each of us has of himself and his position in the social order.[5]

I make these comments to preface the second broad generalization which the social scientist can offer with respect to the effects of housing: the impact of housing varies with the social group that occupies the house. The emphasis in this generalization is not on the manner in which the influence of housing depends upon characteristics of behavior, but rather on the variation in the effects of housing that result from the characteristics of the group which is behaving, regardless of the kind of behavior they exhibit. The last clause in the previous sentence is the important one, for if we reflect on our earlier discussion we realize that social action sometimes is influenced by housing even if the action occurs outside the house and even if the action does not require the physical facilities of the house for its successful completion. I am sure all of us can think of individuals and families who have been made happier, less irritable, or more hopeful in their orientation to life when they moved to a more luxuriant or more handsome section of the metropolitan area or have acquired a more spacious dwelling unit. Changes often have occurred in behavior which clearly had its locus outside the house: the occupant or possessor of a more commodious dwelling was more cheerful in relations with friends in the park, or with colleagues in the office, or with children in school.

How do we reconcile the apparent inconsistency between the claim made in our discussion of the first broad generalization with the present assertion that housing also can influence social action occurring outside the house, including action which does not meet our definition of housing-specific behavior? The answer is to be found in the fact that the house, by virtue of its condition as a possessed object, has a significance in addition to its significance as a physical facility. It also has symbolic meaning. To the occupant himself and to his neighbors and friends, the house is an index of a new class position. Like any other object which is desired and which an individual then acquires, the house can serve as a reward in a means-end scheme of aspiration and achievement. Therefore, possession of a new house can reduce irritability, not because the physical properties of a house—say, greater spaciousness—immediately and directly have a positive effect on mental health through the reduction of overcrowding, but rather because possession and occupancy of new house can be a sign to a family that it has moved upward in the social scale. And upward mobility, we know, can have all sorts of positive psychological influences.

Although our evidence on this matter is very fragmentary, there seems to be considerable variation in the symbolic significance which different groups and individuals in the United States attach to housing. We have said that possessions constitute one criterion which people take into account in evaluating the class positions of others, and in judging their own rank in the social class system; but possessions obviously are not the only element. People also rank themselves and others in terms of performances, like how fast they run, how well they do on examinations, and their productivity on the job. Different social classes and nonmaterial cultures assign different relative importance to possessions, qualities, and performances in evaluating the social position of an individual and family. It also seems to be true that, within the category of possessions, one individual or group will think clothes are most significant, another will regard the brand of car as crucial, while still another will believe the design and cost of the house is the most important possession to evaluate.

I believe that variations in the nature of the possessed objects which are used to rank individuals and to form self-images may help us understand why the relocation of families to public housing appears to have had a more salutary impact on the behavior patterns of slum tenants in American cities thirty years ago than it has had recently. Many of you may be familiar with the findings of social science studies conducted in the United States in the 1930's which indicated that the rate of juvenile delinquency, for example, was reduced when families moved from slums to public housing. Recent studies of similar relocation programs, however, do not report a corresponding decline in the rate of delinquency or in the rates of other forms of deviant behavior.[6] The discrepancy between early and recent findings is often explained by the claim that American social scientists of the 1930's were less careful researchers and less sophisticated methodologists than the behavioral scientists who conduct comparable investigations today; or it is said that the scientists of the past uncovered changes in deviant behavior as a result of rehousing because their liberal ideology led them to expect changes.

An alternative interpretation, which I support, is that the incompatibility between the studies is a consequence of the fact that the inhabitants of American housing projects in the 1930's and the inhabitants of the projects today are samples drawn from two very different populations. During the earlier period, some people with middle-class ideals and values were forced to live in slums because of reverses suffered in the Great Depression. But the economic circumstances in which this group of slum dwellers found themselves did not diminish their commitment to their house as an object according to whose properties they ranked themselves and in terms of which they anticipated others would rank them. When they were relocated to newer, more handsome, better equipped public housing projects, these temporary victims of the Depression saw themselves as acquiring a possession consistent with their middle-class ideals and aspirations. Public housing represented to them a sign of upward class mobility or remobility, strange as this fact may seem given the contemporary attitude in the middle class toward public housing. The tenants of public housing today are drawn from levels of the class structure which are less likely to regard the house as a significant possession influencing social ranking, perhaps because their life history leads them to invest their loyalties in objects which are more easily movable, such as automobiles.

I confess I know of no specific study which bears out this interpretation, in spite of its growing currency, but it would not surprise me to come upon one which confirmed it. Social scientists lack specific investigations which detail the possessions which different classes select in evaluating the social positions of others and in forming their own self-image. However, there are studies from which one might possibly extrapolate supporting materials. For example, we can point to a variety of informed discussions within the recent social science literature concerned with city planning and the role of the physical environment that suggest the existence of considerable individual variation in sensitivity to, and concern for, the physical environment. I am thinking here of the interesting work by Paul Schilder and Harold Searles,[7] two psychoanalysts, who have explored the ways in which individuals make use of the physical environment in developing their sociological self-image; and of Marc Fried's examination of the attachment of some of the ethnic Italians from Boston's West End to the territorial space they had left behind when urban renewal took over.[8] If individuals and ethnic groups differ in the meaning and importance which they attribute to the spatial world, it is not unreasonable to assume that similar variations occur along social class lines.

I would like now to summarize the four major points which I have made and to suggest their implications for housing policy. First, the social effects of housing are best understood if we keep in mind the distinction between material and nonmaterial culture. Second, like every other object in the material culture, the house is embedded in a web of nonmaterial culture; houses are possessions, and, as such, they symbolize and express the class position and value systems of the persons who possess them. Third, housing can affect social action directly because it is an object which can facilitate or thwart social action if the action takes place within the house and if the behavior is housing-specific, that is, requires the properties of the house for its successful completion. And fourth, changes in housing are most likely to influence social action indirectly if the possessors or occupants of the house and their neighbors regard the house as a sign of their social position.

One of the most important implications of these conclusions for housing policy is that many of the social problems of cities and urbanized societies are not amenable to solution through improvements in the properties of the house or dwelling unit.

We often hear it said that rehousing, or the rehabilitation of existing housing, will eliminate deviant behavior, reduce family disorganization, and the like. I suppose the claim is made less often by professional experts than it is by politicians who must appeal to legislatures and to the public for financial support of housing programs. It is an argument which should be resisted.[9] After all, there is little reason to believe that the conditions under which the house as an element in material culture is able to influence behavior really obtain very often with respect to the problems public housing in American cities is called upon to solve. Too much of the troublesome, deviant behavior occurs outside the boundaries of the dwelling unit, too much of it falls short of the criteria which define an act as housing-specific, and a large part of this behavior occurs among social groups to which the house is not a significant possession.

The very limitations that prevail in the capacity of the house to influence social action make it all the more important, however, that housing programs direct their attention to developing innovations which fit the specific conditions under which housing potentially is able to influence behavior. This means that greater ingenuity must be shown in incorporating into the design of dwelling units those physical properties which help to enlarge the range of social action that takes place within the house and that is housing-specific. There is good reason, for example, for persons involved in the planning of government-financed housing projects to take more seriously the modern architectural ideas that propose designs for creating variable spaces. The great advantage of movable walls is that they increase the possibility of using the house and the housing complex for social actions that now must take place elsewhere, actions such as club meetings, study groups, and parties for adults and adolescents. To the degree that people are encouraged to use their dwelling units for activities that now ordinarily meet outside the house, other properties of the house then can begin to have an influence on behavior.

Perhaps even more worthy of exploration is the possibility that housing agencies can profit from an awareness of the conditions under which housing affects action through adopting policies which transform the nonmaterial culture of housing. This already is being done, I suppose. Surely it is recognition of the distinction between the material and the nonmaterial culture of housing and of the need to coordinate their elements more expeditiously which underlies the various health, welfare, and educational programs which now accompany the public housing activities of the Department of Housing and Urban Development in the United States and the Ministry of Housing and Local Government in Great Britain. I wonder, however, whether the content of these programs should not be intensified, particularly in the direction of capitalizing on the desire of all groups in our society to improve the excellence of the objects which they possess. Still more ingenious ways must be provided to enable low-income groups in America and in Europe to own as well as to occupy dwelling units that are physically adequate. Programs should be developed to enable low-income groups to acquire housing in areas of the cities and in the suburbs which already are defined as prestigious. Programs such as these often have been recommended in the literature on housing, and in a few metropolitan areas of the United States they are being tried out on a small scale. Our analysis suggests that the cumulative positive effect of housing in the suburbs or cooperative low-income housing in the cities may be even greater than the new policies have led the housing movement to expect. Privately owned housing may involve the less fortunate groups more fully in the dynamics of the social class system by providing them with the currency, in the form of possessed objects, in terms of which favorable judgments are exchanged in our society.

The analysis of the conditions under which housing can influence social action leads me to recommend, too, that we reshape our housing policy so that the housing needs of the large lower-middle class in the United States no longer are overlooked. I was encouraged that the Secretary of Housing and Urban Development and President Johnson agreed on the need to request from Congress the statutory power to inaugurate suburban development planning oriented toward the establish-

ROBERT GUTMAN

ment of new centers of settlement. One of the major burdens of this discussion has been to suggest that the restrictions surrounding the capacity of housing to influence behavior in a positive way are very severe and that housing is more likely to influence action indirectly through its symbolic significance. Yet the sad fact is that housing policy for too long has been aimed to meet the needs of that segment of the American population which is least likely to recognize the symbolic value of housing, while the middle class, which already views a house as an important possession, is left to satisfy its demand through the brutal and impersonal processes of the housing market.

## Notes

1. Herbert Gans, *The Levittowners* (New York: Pantheon Books, 1967), Chapters 3, 8, and 11, *passim.*

2. Bennett Berger, *Working Class Suburb* (Berkeley: University of California Press, 1960).

3. A popular version of this idea is presented in Philip Wylie, *Generation of Vipers* (New York: Rinehart and Co., 1942). The theory underlying the association between elements of the material culture and some but not other elements of the nonmaterial culture is given in Julian Steward, *Theory of Culture Change* (Urbana, Ill.: University of Illinois Press, 1955).

4. Daniel M. Wilner, et al., *The Housing Environment and Family Life* (Baltimore: The Johns Hopkins University Press, 1962). An excellent review of the literature dealing with the effects of rehousing on deviant behavior can be found in Alvin L. Shorr, *Slums and Social Insecurity* (Washington: U.S. Department of Health, Education, and Welfare, Social Security Administration, Division of Statistics, Research Report No. 1, 1963).

5. The significance of possessions in determining social rank, along with the role of other criteria, is discussed in Talcott Parsons, "A Revised Analytical Approach to the Theory of Social Stratification," in Reinhard Bendix and S. M. Lipset, eds., *Class, Status, and Power* (Glencoe, Ill.: The Free Press, 1953), pp. 92–128.

6. Schorr, *op. cit.,* Appendix A.

7. Harold F. Searles, *The Nonhuman Environment in Normal Development and in Schizophrenia* (New York: International Universities Press, 1960); Paul Schilder, *The Image and Appearance of the Human Body* (London: Kegan Paul, Trench, Trubner, Trubner and Co., 1935).

8. Marc Fried and Peggy Gleicher, "Some Sources of Residential Satisfaction in an Urban Slum," *Journal of the American Institute of Planners*, XXVII, No. 4 (1961), 305–315.

9. John P. Dean, "The Myths of Housing Reform," *American Sociological Review*, XIV (1949), 281–289.

# 11 THE MAYOR

## Richard C. Lee

It should be easy for me to describe the position of mayor, since I have been a mayor for fifteen years. However, the job of running a city has become such a large part of my life that it is difficult to describe in a short statement what a mayor is, what a mayor does, or even what a mayor should be. First, however, let me give something of my own background. After several years as an alderman, which is the term for a City Council member in New Haven, and a tour of duty as Press Director for Yale University, I ran twice for mayor of New Haven before being successful in 1953. I have now served seven consecutive, two-year terms and expect to run again for an eighth term.

In this brief discussion I wish only to strike a note in your minds and in your hearts, so that you will have a better understanding of the increasing significance not only of the leadership roles in your own community, but, indeed, of the problems faced by cities all over the world. I am not a scholar, nor am I an expert in urban history, but I have spent fifteen years running, supervising, leading, cajoling, and sometimes badgering a city and its people; and it is from this experience that I write.

New Haven, Connecticut is a city of 140,000 people which was founded in 1638. The land area of the city is 21 square miles, and the first settlements were clustered around an excellent harbor on Long Island Sound. The central part of the city is well known for its plan of nine equal squares, centered around an open green. The mayor of New Haven, as in most cities in the United States, is the popularly elected executive head of the city. The mayor's powers may differ from city to city, but in general he has legal control over the more important community affairs. He is, for example, the chief financial officer and the person in charge of future physical development, as well as the person responsible for community services such as police and fire and libraries.

The mayor is, as well, the spokesman for the city and its contact with state and federal officials. In America, cities are created by the states, and nearly all their powers, such as the power to tax and to legislate, are derived from and defined by state legislation. However, in recent years as cities and their problems have grown rapidly, the federal government has offered direct assistance to individual cities. This new federal role is partly due to the size of the urban problem and the large sums of money required for comprehensive solutions and partly due to the fact that the state governments have been slow to recognize the financial needs of cities. This trend is changing, and state governments are assuming more and more financial responsibility for urban programs.

But running a city requires more than just money or administrative skill. The mayor must be more than a budget director or administrator of the city's physical growth and the like. A mayor is also the leader of civic spirit, the catalyst, the chief complaint officer, and the inspiration to all the varied people who live in a city. In America a mayor is elected for only a two- or four-year term. Therefore, perhaps his most important task is to produce results from promises in a brief time. This is no easy task, and the mayor is constantly under pressure to make programs come alive as quickly as possible. The mayor, unlike state or national elected officials, is always on the public firing line. He lives with his family, attends church, and walks the streets in his jurisdiction. Every citizen knows the mayor and looks to him for leadership. It is his responsibility to think up new ideas, to propose and support legislation based on these ideas, and then to find ways to pay for the new programs. Urban centers are the laboratories for social and economic theories, which through legislation result in programs. The mayor is at once the judge, the jury, the defender, and the prosecutor of these programs. The mayor must make the programs come to life and reach as many of the intended people as possible. Not only is the mayor charged with the proper execution of programs, he is also charged with knowing the feelings, the thoughts, the problems, the complaints, and the plight and poverty of his people. He is charged with changing programs in midstream if they are not reaching the goals for which they were written. He is charged with telling the national and state governments that programs are obsolete or that new programs are needed.

In New Haven, as in a number of American cities, physical and human renewal programs have been underway for nearly a decade. These programs are still new, and even when they first began they were long overdue. The entire world has been slow to realize the implications of urbanization, and much of what has been done in the last several years has been barely enough to keep pace with changing conditions. However, recent attention to the plight of cities has also focused attention on the men in city hall—and specifically the mayor. At long last, the caliber of men filling posts of mayor all over America, and seeking the post of mayor, indicate that people have finally begun to recognize that the key unit of our society is the city, and that the responsibility for leadership in a city rests with the mayor. The mayor in America today has a major role in shaping and molding urban life. He is no longer a specialist in attending wakes and funerals, and he is no longer the familiar figure with a potbelly who smokes cigars and exhorts with pomp and rhetoric.

Today the mayor must have the executive ability and administrative skills of a corporation president; and he must manage a multimillion-dollar budget. He must know the business of all who trade in a city, in order that he may bargain and negotiate effectively for the public good. And he must bring together an intelligent, dedicated staff of professionals who can give creative, imaginative treatment to cold legislative programs. The mayor must prod the people of his community and his staff into action. He must set forth his own goals. He must build the spirit of the community. And he must lead his community into an understanding of what one side wants, what the other side desires, and what each will accept. And while the programs ferment, and while the intellecutal or the educator proclaims disagreement or criticizes, the mayor must also face the wrath, the doubting, and the criticism of those to whom every two years he turns for support and for renewal of his responsibility. Alas, unlike some of his critics, he does not have job tenure. What are our urban problems? What are the goals and the tools we have for combatting these problems? I will answer these questions with a brief historical account:

In this century the wave of urbanization has occurred in nearly every nation. Wherever this locational change has taken place, it has created social changes which can cause problems as well as opportunities. In America, during the 1930's, the New Deal of President Roosevelt created the vision of a new America and pledged "freedom from want" for the one-third of America's people who were ill-housed, ill-clothed, and ill-fed. This pledge had an air of urgency and immediacy in those desperate days of the Great Depression. It promised employment, better housing, and social security for the elderly. But the New Deal was in great part aimed at the rural poor.

Only in the last decade—perhaps only in the last five years —has modern society, not just in the United States but all over the world, begun to recognize the critical importance and significance of cities, their needs and their problems. At last the voice of the cities is now heard in the land, and the men who run the cities are becoming known as truly the men for all seasons. We now have had "The Great Society" of Presidents Kennedy and Johnson, and the massive "War on Poverty." And what was their reason for being? These programs came into being because after thirty years the problems of disadvantaged Americans were the same. However, significant change has occurred since the 1930's, for now the overwhelming majority of these people have migrated to and live in cities.

As a mayor, I know that attitudes of indifference, lack of concern and apathy, still exist toward these urban problems. But I also know and firmly believe the city should be the finest expression of man's activities, his ideals, and his beliefs. I believe the city should be the finest stage for culture, education, and the arts. I believe the city should be the marketplace where all people meet on equal terms—to trade, to live, to work, and to dream. I believe a city is the highest expression of civilization. Its buildings express the wonder of religious belief, the efficiency and sometimes the greed of business, the mobility of modern man, and the deep roots of history. I believe, finally, that a city is first and foremost its people. It is a place where people are born, where people live, where people are educated, and where people die.

The focal point of urban life with all its chaos, excitement, and controversy is the mayor. He is at once both an old and a new type of mayor, for not only is he the chief complaint officer, the ceremonial head, and the man who shares the sorrows and despairs of his people, he is, as well, the captain, the pilot, the maker of policy, the administrator, the planner, and indeed, the reflection of all that is both good and evil in the people he leads and the city he guides.

The mayor and the city have come of age, because cities are where everything is happening, both good and bad. Men of great ability have finally begun to accept the challenge not only of the city, but of the city hall. The new look is burgeoning across the nation; the mayors are more often like John Lindsay of New York and Ivan Allen of Atlanta than they are like the ceremonial figures of the first half of the century.

Mayors are concerned with the direction of their city, and New Haven is a good example. But New Haven is not unique. New Haven has many of the same problems that plague New York, London, Rome, or Tokyo. The differences are only of geography, magnitude, and degree. The nature of the problems and their existence are the same for nearly all the urbanized world. My job as a mayor is to make people aware that the major domestic problems begin and end in the cities. My plea is for understanding and support from all who can help as we seek answers to the solutions of these problems. My plea is for understanding of the problems of people in cities. I have found that being a mayor is a most rewarding public office. It combines the professions of law, business, architecture, and planning into an entirely new profession—the profession of the city builder. The city builder is concerned not only with bricks and mortar but, as well, with the spirit and the souls of the impoverished people, their hopes, and their opportunity.

I firmly believe that the position of mayor at this point in history is the most vital and even the most important public office in a democratic government. For all who seek to improve society look to the cities and to the mayors for new ideas and the ability to carry out existing programs. Running a city is a challenge to the executive ability of any man, and working with all the people who live in a city is perhaps the most rewarding part of the job.

I have devoted a good portion of my life to the pursuit of excellence in urban America. It can be said, perhaps, that my entire adult life has been directed to cities, for when I was not running for mayor, I was preparing for the job.

Some people say to me, what have you done in life? Why haven't you been more than just a mayor? My answer is that I hope I have been more than just a mayor. I hope I have helped to begin to rebuild and to develop the human resources of a city—a special city which is the home of Yale University, one of the great educational institutions of our time, a city I love and a city where I was born and where I hope to spend the rest of my life.

I will conclude with a profound piece of fictionalized history which I think describes very well the feeling and commitment which I feel as a mayor. In Edwin O'Connor's novel, *The Last Hurrah,* the hero is a mayor named Skeffington. In the final scene the mayor is dying and, as a Catholic, has been given the last rites of his church. He was a rogue when he lived, but a lovable, likable, and understandable rogue. At the bedside was the pious, unctuous father-in-law of the mayor's favorite nephew who, even as life was ebbing from Skeffington, looked down at him and said:

Well, no matter what some of us may have thought of the past, it is all different now. We know he has made his peace with God, and I think we can say this—knowing what he knows now, if he had to live his life over again, there is not the slightest doubt that he would do it all very, very differently.

The dying old man on the bed stirred, his eyes opened, he raised himself slightly and, in those dying eyes, there was again the old challenging, mocking gleam; and, taking charge for the last time, he bellowed:

"The Hell I would!"

This is the way I feel about my job as mayor. I'd do it all over again.

# 12 CRIME

James Q. Wilson

In the United States, as in many countries in all parts of the world, there is a widespread and growing concern about crime. Not every country perhaps is experiencing rising crime rates, but almost every nation which is undergoing rapid urbanization and industrialization or which is attempting to produce a common culture or a single national identity out of various ethnic and religious elements must now face the realization that this progress, like all forms of social change, has a price. That price is the temporary—and sometimes permanent—dissolution of old social ties, the collapse of ancient familial controls, the prominent display of new and exciting modes of behavior, and the increased conversion into necessities of those material goods once thought to be luxuries.

In America, as I suppose in most countries, whatever their political arrangements or social philosophies, one rough measure of the progress we have made in increasing the prosperity of the masses and the liberties of the individual has been the higher rates of certain kinds of crimes, especially crimes against property. The more people can buy, it seems, the more they also steal. By contrast, the general improvement in the standard of living has tended to reduce certain kinds of crimes against the person, such as murder. As people become more prosperous they become more civil but no less larcenous.

These, at least, are the appearances. Perhaps because they seem so paradoxical—perhaps because we find it hard to believe that a rising standard of living should be accompanied by a rising rate of theft—these appearances have been challenged by various scholars. Some have denied that crime rates are in fact going up.

Others have argued that our criminal statistics leave so much to be desired that we cannot tell whether crime is increasing, decreasing, or staying the same. And there are great weaknesses in our data—as in most countries' data—on crime. Many crimes that occur are not reported by the victim to the police; some police departments do not keep accurate records of those crimes that are reported to them; the legal classification of crimes varies from state to state and, within one state, from time to time. When New York City, for example, changed its police reporting system, the apparent number of robberies increased by 400 per cent in one year. Finally, such crime statistics as we have—at least for the nation as a whole—are of little value for periods earlier than about ten years ago; if they are correct in asserting a generally increasing crime rate, that trend may only be a temporary phenomenon and quite unrelated to the long-term crime trend for, say, the last hundred years.

Partly to find answers to such questions, but mostly to provide some visible response to the widespread popular belief that whatever the scholars may say crime is bad and getting worse, President Johnson created a Commission on Law Enforcement and Administration of Justice and supplied it with a large staff of researchers and consultants. It issued a report—one summary volume and nine task force reports on special subjects—that contained over two hundred recommendations and, what is important, a good deal of new information.

One of the Commission's major contributions to our knowledge of crime was a survey of ten thousand households selected at random from all over the country. Persons at these addresses were interviewed to discover whether during the previous year they had been the victim of a crime and, if so, what they and the police had done about it. This survey was supplemented by other, similar studies in certain high-crime-rate cities. The results showed that, as everyone suspected, there was a great deal of unreported crime in America—twice as many serious assaults, three times as many burglaries, four times as many rapes as the police figures had indicated. But the survey also suggested that most of this hidden crime was the result, not of the police concealing crimes known to them, but of individuals failing to report them to the police at all. And their reasons for not reporting them were quite understandable and probably would operate in almost any country. Since most assaults involve people known or related to each other—husbands and wives, estranged lovers, drinking companions—many assaults were not reported because the victim did not want to get the assailant in trouble with the police or, in other words, did not wish to get himself in further trouble with his assailant. Most burglaries, since they are crimes of stealth, are never solved by the police here or anywhere else. Knowing this, the average citizen often feels that, since the police are not likely to get the stolen property back, reporting the crime is not worth the trouble.

What this survey could not show, of course, is whether crime is increasing by as much as the official figures suggest. To answer that question would require having an earlier survey against which the present one could be compared. The Commission's experts did the best they could with the official figures and found that no easy answer is possible. One of the central problems concerns the meaning of the term "crime rate." Technically, it refers to the number of crimes committed per one hundred thousand population. If the crime rate goes up, as it has, some people assume that this means that the average man is more criminal today than he was ten or twenty years ago. But that is not necessarily the case, for not all groups in the population are equally likely to commit crimes. Arrest records suggest—but cannot prove, since those arrested may be unrepresentative of those who commit crimes—that men commit more crimes than women, that young men commit more crimes than older men, that Negroes commit more than whites, and that big-city dwellers commit more than farmers or small-town residents. Thus, if the total population changes its composition and consists of higher percentages of young Negro males living in big cities, the total crime rate will almost certainly go up even though the average person (or the average Negro, or the average young male) is no more criminal today than ten years ago; indeed, he may be a good deal less criminal.

Precisely these changes have been occurring in American society. People of all races have left the farms for the cities, where there are more things stolen because, among other reasons, there are more things to steal. The Negro proportion of the total population, but especially of the big-city population, is steadily rising. There are more young people than ever before—beginning in 1961, nearly a million more youths have reached the age of maximum criminal risk each year than did so the prior year. Taken together, these changes in our population explain, in the Commission's opinion, about half the total increase in the crime rate that has occurred since 1960.

What can explain the other half? Partly, no doubt, it reflects an increased accuracy by the police in keeping records. The big-city police departments which have made significant changes in their crime reporting systems over the last eight years account for nearly one-fourth of all serious crimes against the person and about one-sixth of all serious crimes against property. Partly it is because people who once committed crimes in areas where such incidents were not recorded now live in areas where they are. Negroes living in the rural South, for example, have been subjected to a double standard of justice—one Negro stealing from or assaulting another Negro (and most serious crime occurs among members of the same race) was often ignored by law enforcement officials. Either it was not even recorded as a crime or, if recorded, little was done about it. A white assaulting another white was treated more seriously and, of course, a Negro assaulting a white was treated the most seriously of all. Now, Negroes tend to live in big cities where more professionalized police departments count all crimes reported to them no matter who commits them. Another reason for the rise in crime is that many people, having achieved middle-class status, now regard (and report) as a crime what they once treated as an insult and handled by private vengeance. An unemployed, uneducated man living in a tough neighborhood may return a punch in the nose with another punch in the nose; a respectable, middle-class man living in a nice suburb may respond to a similar punch by calling the police and demanding that the assailant be arrested. Finally, a prosperous society displays a great deal of attractive merchandise on the shelves of its stores and, with rising labor costs, often in stores where one serves himself rather than is waited on by a clerk. With so much more unguarded merchandise lying about, it is not surprising that some Americans have come to take the invitation, "serve yourself," too literally. Shoplifting, accordingly, has risen dramatically.

These factors have affected many countries, not just the United States. Crimes against property seem to be on the increase in all nations experiencing sudden bursts of prosperity. Since 1955, reported thefts have more than tripled in West Germany, the Netherlands, Sweden, and Finland and more than doubled in France, England, Wales, Italy, and Norway. Crimes against the person have not increased so fast, however, and in some cases have actually declined. They went down in Norway, for example, but went up—dramatically—in England.

It is, I imagine, these crimes of violence which most interest a foreigner studying the United States, just as they most interest an American concerned about his safety on the streets. The murder rate, for example, is known with a fair degree of accuracy, and though it has not increased, and in fact may have declined slightly over the past ten or twenty years, it is still high by the standards of most European nations. In 1962, the murder rate in the United States was three to four times higher than in France, Germany, Ireland, Great Britain, Canada, or Japan. And, of course, one special form of murder—the so-called gangland killing—is probably of special interest to persons from countries which are free of organized or syndicate crime. Since the second decade of this century, there have been about one thousand unsolved gang-style murders in Chicago alone. Small wonder that when a foreign visitor thinks of Chicago, he often thinks of a machine gun.

In fact, gangland killings, serious as they may be, are only a tiny fraction of all murders. If organized crime did not exist, the murder rate would still be about what it is today. What, then, accounts for this degree of violence? No one is quite sure. We do know that murder is more common in the southern part of the United States than in the northern or western parts. This may be the result of the strong tradition of personal honor—and its corollary, personal vengeance—that is found there. It may also be because the South is poorer and more rural; the murder rate in the South is slowly going down as the South becomes in many respects more like the rest of the country— urbanized, industrialized, prosperous. But this cannot explain the whole difference. Vermont, for example, is a rural state, just as is Georgia; yet in the past, Georgia's murder rate was more than twenty times higher than that of Vermont. (By contrast, Vermont's suicide rate was almost twice that of Georgia.) And this is not simply because Georgia has more Negroes than Vermont— the rate at which whites are killed by whites in Georgia is over five times that at which whites kill whites in Vermont. We know very little about these regional differences, but it is striking that in many countries—one thinks of Italy, for example—the violence of the southern region is legendary.

Another explanation for the comparatively high level of violence in the United States is to be found in the kind of society we have tried to create. This country is a nation of immigrants from all parts of the world. There is scarcely a religion, an ethnic group, or a nationality that is not represented in large numbers in the American population. A principal problem of American government has always been to reconcile the competing demands and traditions of groups with distinctive life styles. In this country, our large cities have been the principal locus of such efforts. Immigrants had not only to endure the strain of a long and uncertain journey but to confront at journey's end the bewildering and unfamiliar customs of a new world amidst the buzzing confusion of a large city. The shock of transplantation was great, and perhaps the suddening weakening of old social and familial ties gave rise to an increase in crime.

During the last decades of the nineteenth century and the first decades of this one, many books and articles were written commenting on the disproportionally high crime rates among certain immigrant groups. Their alleged proclivity to crime became one reason used by some to argue for a restriction on the entry of immigrants into this country. In 1931, the subject was still so important as to lead a presidential commission— the so-called Wickersham Commission—to devote much of its report on law enforcement to a consideration of the criminal tendencies of the foreign-born. Interestingly enough, it found no evidence that the immigrants, taken as a whole, had crime or arrest rates any higher than the native population; indeed, it found considerable evidence that their crime rates were in fact lower. But this was only an average; groups from certain countries—typically, the newest and poorest arrivals—had crime rates higher than the average, while other immigrants—often, the older and more well-organized settlers—had lower rates.

The immigration to our big cities continues today, but the present migrants are less the foreign-born than the rural Americans, white and Negro, who have left the farm to come to the city. In 1910 less than a third of all Negroes lived in the large metropolitan areas; today, over two-thirds do. These migrants face all the problems of previous groups, and additional ones as well. They are marked by a badge of color which unfairly makes the law-abiding Negro bear the stigma of his lawless brother. The migration was not only from the farm, but from a system of life originally formed by slavery and perpetuated by agricultural tenancy in which property ownership was difficult, the family disorganized, and free access to the public institutions of the region almost nonexistent. When the typical European left his homeland to come to this country's cities, he was a peasant; when the Negro left to come here, he was part of an agricultural proletariat. The European brought a whole culture with him, while the Negro could bring only a way of life formed in great part by his exclusion from the dominant culture.

The results of these waves of migration can be seen in the careful studies made in Chicago for over thirty years by Clifford Shaw and Henry McKay. They found that the areas of that city with the highest rates of crime and delinquency were those near the center where incomes were the lowest, housing the most deteriorated, families the most disorganized, truancy the greatest, and neighborhood institutions the least developed. And most importantly, the prevalence of high crime rates persisted in these areas over time, even though the people living there changed. Native-born whites were replaced by the Irish, the Irish by Italians, the Italians by Jews, the Jews by Negroes—and throughout, crime rates remained high (though not as high for every group).

146

No one is quite certain as to the social processes which have kept the crime rate high in these areas. It is not, we know, simply the fact that newcomers are steadily arriving in the city. At least two studies have shown that, among Negroes at least, those born and raised in the big city have much higher rates of delinquency than those who have recently migrated to it. Apparently, it takes time before the new arrival either learns enough about the city, or is sufficiently frustrated by its problems, to take up crime that is serious enough to attract the attention of the authorities. No matter how long it may take, however, the high crime areas eventually affect the migrant, and this is especially true if the area is changing in population rapidly. Of the seventy-four neighborhoods which make up the city of Chicago, the crime rates have been fairly stable in fifty-nine but have changed dramatically in fifteen. Those where the rates increased were ones in which significant population changes occurred—in each case the movement into the neighborhood of a new ethnic group and the retreat from there of the old. Those where the rates decreased were ones in which the population, after a change several decades ago, had now become fairly stable with little turnover and with a more or less homogeneous character. Those areas where the decrease in rates was greatest were the areas where, thirty years ago, the rates had been the highest. But most interestingly both the areas with the greatest increase and the one with the greatest decrease are predominantly Negro. Thus, it may be—no one knows for certain—that it is migration, disorder, and change involving low-income groups that raises rates, and stability and homogeneity (even though incomes may remain low) which contribute to a lowering of rates. (It should be borne in mind, of course, that even in low-income areas where the rates have declined, they are still much higher than the city average.)

The net effect on crime of the social changes now under way in America—steadily increasing prosperity, new patterns of internal migration, the reduction in foreign immigration, the movement of the middle classes out of the central cities and into the suburbs, the elimination of great disparities in education—is hard to assess. Very few studies of crime over several decades

147

have been attempted, and those that have been done are based on imperfect information. There have been two or three efforts to measure criminal patterns in Boston, Chicago, New York, and Philadelphia for periods ranging from twenty to a hundred years. In general, these studies seem to agree that the crime about which we have the most reliable data—murder—has been declining more or less steadily though erratically for a century or more. Manslaughter, by contrast, has increased greatly—owing, no doubt, to the introduction of the automobile. About burglary and robbery we can say rather little, since the accuracy with which these crimes are reported has varied so much. Over shorter periods, however, some patterns do emerge—an economic depression apparently increases the number of burglaries, a war decreases them. The reasons are not hard to imagine. A depression leads people to steal what they can no longer buy; wars, on the other hand, not only bring generally high incomes, they take out of the population that group—the young men—who are most likely to steal and place them in uniform.

Thus, in the United States, as in many countries, there are partially offsetting trends at work. Long-term economic growth and the acculturation of ethnic minorities are working to reduce the volume of serious crime; the younger age and increased urbanization of the population, coupled with the widespread use of the automobile, have worked to increase it. A major, and for now unanswerable, question is whether in addition there has occurred a change in the customs, morals, and values of the society that will have a long-term effect on crime. Again, two possibilities can be envisaged and may indeed be at work simultaneously. One is that the increased civility of society, arising from the entry of more and more people into middle-class status, will make people less tolerant of crime and more inclined to invoke law enforcement agencies to cope with it. These higher standards of civility will, the theory goes, result in less actual crime being committed but (perhaps) in more crime being reported as people raise their standards of appropriate behavior and lose their toleration of disorder, and thus treat as criminal that behavior once regarded as simply boisterous. The opposite theory is that the very prosperity of society has led to the breakup of the extended family, a weakening of parental controls over youth, a secularization of religion, and an increased toleration for licentiousness that will ultimately produce more crime, especially juvenile crime. In short, one theory anticipates that an affluent, industrial democracy will be more respectful of the rights and privacy of others and less tolerant of public commotion; the other theory says that the price of these gains will be a decay in moral values and a weakening of the spiritual and ethical bases of organized society.

Each theory leads to certain policy proposals. Those who see American society in terms of the spread of middle-class morality fear not only the conformity they believe this will induce, but the intolerance of the middle classes toward the deviant but not harmful behavior of the lower classes. Accordingly, there is a significant and growing body of liberal and radical sentiment in this country that resists the imposition of what it terms "middle-class standards" on the poor and argues for a change in police practices to end the "harassment" of lower-income groups and ethnic minorities, a change in court and correctional institutions that will recognize the different standards of various subcultures, and a change in our laws that will permit certain practices—homosexuality, the taking of certain kinds of drugs, and the frank display of sexual activity—that are now illegal. Those, on the other hand, who see society in terms of a decay in moral and familial standards argue that misconduct cannot be tolerated whatever its subcultural origins, especially when lesser forms of theft and disorder can lead (if ignored) both to more serious crimes and to a generalized contempt for law enforcement and a resort to private vengeance. In this view, the efficiency of and support for the police should be increased, familial responsibility over the young should be strengthened, and public immorality should be curtailed because, though it may hurt no one in the same sense that robbery does, it encourages licentiousness and creates contempt for decency.

I would imagine that these arguments are not unfamiliar to people in every country in the world. Whatever else may divide us, a willingness to conduct passionate and interminable arguments about public morality is not one of them. Fortunately,

149

the factual premises of either moral argument cannot be confirmed by contemporary scholarship, and thus professors need not take sides. (They need not, but of course they take sides anyway.) There is one thing that scholars have learned, however, which is relevant to the issue. The effects of crime, whatever its causes, are not equally distributed throughout society. Some people suffer more—that is, are more likely to be the victims of crime—than others. The survey of the victims of crime conducted for the president's Crime Commission showed that Negroes are far more likely than whites to be the victims of murder, rape, robbery, assault, burglary, and auto theft and that lower-income persons are more likely than those of higher income to be the victims of murder, rape, robbery, and burglary. Only for auto theft and larceny are upper-income persons victimized at about the same rate as the poor. Also, studies of victim-offender relations in specific crimes—principally murder, rape, and assault—show that the victim is typically of the same race and about the same social class as the offender.

Such findings should come as no surprise to those who have experienced life in the inner parts of the larger cities in almost any country. The congested, low-income, transient central neighborhoods are those where both criminals and their victims reside. The apparent increase in big-city crime rates in America over the last five or ten years has occurred primarily at the expense of the poor and the ethnic minorities and has led Negroes as well as whites to regard "crime in the streets" and the need for police protection as an important political issue. It has had another effect as well. The continuing movement of middle-class families out of the central cities and into the suburbs has been hastened, if not in part caused by their concern over central city crime. Thus, whatever else may be the result of the increasingly middle-class character of American society, it has made the average citizen less willing to tolerate central city crime rates and more able to do something about it—namely, move out. Even though our big cities today are probably much less crime-ridden than they were in the second half of the nineteenth century, the currently rising crime rates are having an effect they did not have a hundred years ago—they, along with other central city problems such as poor schools, high taxes, and traffic congestion, are working a permanent change in the character and social composition of the cities that will have profound but hard to judge effects on the economy, government, and cultural life of these cities for many decades to come.

# 13 MIGRATION TO AMERICAN CITIES

## Charles Tilly

Since they invented urban life eighty centuries ago, men have been moving incessantly to cities, and from city to city. Times of great urbanization are times of mass migration. The migration of our own time, which is urbanizing the entire world, far surpasses that of all previous ages in numbers of men and in geographic scope. The most spectacular surges to the city are now occurring in Asia, Africa, and South America. But even in North America, after a hundred years of irresistible urbanization, the bustle of migration continues. Americans move, and move, and move. For a century or so, about one American in five has moved every year. Much of the novelty of California reflects the fact that over two-thirds of its huge adult population has come from other states. Migration is creating new societies in North America as well as elsewhere.

North American migration differs from the Asian or African varieties because so many of the newcomers to any American city have spent so much of their lives in and around other cities, or at least in close touch with them. Practically no migrant to Cincinnati or San Francisco faces the kind of change experienced by a Nubian newcomer to Cairo or a Kohistani just arrived in Karachi. With increasing exchanges of inhabitants among cities and a dwindling share of the total population in rural areas, the great majority of migrants to most American cities now come from other urban areas.

Migration itself drained the rural areas and fed the cities. Up to 1910, the number of people in American agriculture actually grew steadily as the area of settled farms expanded. From 1910 to 1940 it remained fairly constant. Since 1940, the far-reaching mechanization of farms, the squeezing out of small producers, and mass departures of tenant farmers from the land have cut the farm population by two-thirds. By 1967, barely eleven million people—less than six out of every hundred Americans—still lived on farms. Even when the number of people on farms was growing or constant, those people had children at a rate fast enough to produce a substantial surplus for migration to the city. When the numbers of farms and farmers began shrinking around World War II, that only speeded up the exodus.

Combined with mass migration from overseas and natural increase in the cities, this unremitting movement off the farm helped American cities grow much faster than the countryside. The cities' share of the total population went from a fifth in 1860 to almost a half in 1910, to around three-quarters today. Most of the rest live in small towns rather than on farms. Will the vanishing of the rural population mean the disappearance of migration? Not at all. Movement of people among cities will surely continue, and may well accelerate. Yet it does mean that the very triumph of the city will eliminate something which has been part of urban life since cities began: the attraction of the farm boy to the metropolis, and his transformation into an urbanite.

The geographic pattern of migration within the United States looks like a Persian rug: some well-defined main lines, some dominant colors, but within them intricate swirls, contrasts, and interweavings. Until a few decades ago, we could have described the chief flows of migrants within the United States as two or three well-defined streams going from south to north and two or three others going from east to west. Even then it would have been important to remember that every one of these streams was the net effect of a very large movement of migrants in one direction and a smaller but still substantial movement back in the opposite direction.

Nowadays, the map of migration flows is changing in two ways: first, more and more migrants are leaping hundreds or thousands of miles from city to city without regard for the old, established paths of migration; second, while the aggregate movement toward the north and (especially) toward the west is continuing, migration is actually scooping up the people in the interior of the United States and throwing them out toward its edges. As oddly as the fact fits the conventional picture of Americans as dwellers in great plains, two Americans out of every five live in metropolitan areas touching deep water: the two oceans, the Great Lakes, or the Gulf of Mexico. The great majority live within a hundred miles of deep water. And that deep-water band is gaining population far faster than the rest of the country. The rapid rise of California, now the most populous of all the states, sums up the movement toward the west and toward the sea.

Why do Americans move around the way they do? Migration depends on three factors: opportunity, information, and cost. The greater the opportunities elsewhere and the greater the flow of information about opportunities, the greater the migration; the higher the cost of mobility, the less the migration. At the broadest level, this comes to saying that improving communication, increasing case of travel, and the growth of a diversified national labor market promote rising mobility. More narrowly, it amounts to the observation that the closer two places are together, the more extensive the existing contacts between them, and the greater the difference in opportunities between them, the heavier is the flow of migrants between them. When it comes to the individual, the probability that he will migrate to any particular place (or that he will migrate at all) depends on the fit between his needs or qualifications and the opportunities available in that place, the channels of communication he has with that place, and the ties or investments he has in his present location.

Most of these statements would be little more than truisms if it were not for the fact that occupational information, opportunity, and cost by themselves account for much of the American pattern of internal migration. Job opportunities produce the main flows of migrants in most parts of the world; they certainly do so in the United States. Whatever else may influence their decisions to move, the overwhelming majority of families migrate when the breadwinner is taking a new job or looking for work. When pollsters ask newcomers to a city why they came, the newcomers usually answer in terms of jobs. Long-distance migration normally speeds up in times of economic expansion and slows down in times of economic contraction. And we can estimate the flows of migrants among American cities with remarkable accuracy simply by using information about local levels of employment.

This does not mean everyone behaves the same way. Precisely because groups and individuals vary in terms of their access to information about distant jobs, the opportunities actually open to them, and the costs of leaving their present homes, their patterns of migration vary as well. Young Americans just leaving school, for example, ordinarily have up-to-date skills, a wider than usual knowledge of job opportunities, and few ties to hold them in place. Predictably, they have especially high rates of long-distance migration.

Compared with whites, American Negroes and Indians start out with less education on the average, face discrimination in hiring, suffer greater unemployment and less job stability, and get different, inferior information about available jobs. As a result, the pattern of migration varies considerably by race. Again, men with plenty of education and technical skill, like engineers or economists, are more likely than other people to be in demand for distant jobs, more likely to belong to national professional networks which distribute information about jobs and job candidates, more likely to be able to quickly reorganize their lives in a new city; and in fact long-distance migration is much more common among highly educated and skilled Americans than among the general population. Finally, a person's present location affects the cost and the attraction of moving somewhere else; depressed areas lose migrants heavily, and prospering areas gain them. So a man's age, race, skill, and present location all profoundly affect the chances that he will move—and if so, how and where.

Because of all this, migrants (especially long-distance migrants) are superior to the general population in education, skill, and occupational desirability. They come disproportionately from the age groups with the most vigor and the highest education—the late teens and early twenties. Migrants to cities coming from other cities or from small towns tend to be above the average in education and occupational skill in the communities they leave behind. They even tend to rank higher in education and occupation than the population already in the city to which they move. People moving off farms are a little different. They are not consistently better off than the people they leave behind; both the least and the most educated predominate in the younger ages, the least educated in the older ones. They tend to be even younger than other migrants, and they are on the whole below the standard levels of education and occupational skill for the city's population. But migrants from farms are only a small fraction of the people coming to any particular city; in recent years, their arrival has not deeply affected the urban population's level of qualifications.

Along with the definite statistical differences in migration by age, sex, race, skill, and present location come differences in the experience of migration itself. There is a great gap, for example, between the migrating business executive and the migrating day laborer. The executive has a job in the new city nailed down well in advance. Often he is transferring from one office to another within the same firm. He has plenty of experience with organizations and cities, which makes settling into the new place easy. We might call this a cosmopolitan style of migration.

The day laborer, on the other hand, rarely has a definite offer of a job before he comes. He does not usually move as far as the executive. It is much more likely that friends or relatives told him about general chances for employment in the new city, and that he came to that particular city especially because there were already friends or relatives there to help him out. They do help him—with housing, work, and getting around the city—and they form the nucleus of his social life long after his arrival. We might call this a local style of migration. The distinction between cosmopolitan and local styles of migration is quite a general one. In America, the local style is more common among people with less education, members of racial minorities, and people from rural areas. It promotes the formation inside the big city of little villages of persons linked by kinship or common origin.

156

Of course, these two ways of organizing migration exist throughout the world. In most countries outside Europe and North America, some variant of the local pattern predominates. In those countries, the expanding cities have often grown out from a compact, western-style nucleus organized around trade or colonial administration, heavily populated by foreigners, and segregated from the native population. Their growth through the local style of migration helps produce cities grouping numerous transplanted villages around a cosmopolitan core.

In the United States, the cities have commonly grown out from nuclei of trade or manufacturing without the same sort of distinction between natives and colonizers. They have had arrangements of transportation and communication making access to the center from the less crowded spaces at the periphery quick and easy. The growth of American cities through the cosmopolitan style of migration combines with these other factors to produce cities grouping cosmopolitan suburbs around a core containing what villagers there are.

This geographic arrangement of American cities shaped a phenomenon which fascinated urban sociologists for years: the fairly regular sequence by which an underprivileged group of migrants first settled together in the downtown areas of high congestion and rundown but cheap housing, then moved step by step away from the center as its members got better jobs and higher incomes, and mixed with the rest of the population in the process. It happened to many groups of immigrants from Europe: Greeks, Italians, Jews, Poles. It is still happening to North American migrants: Puerto Ricans, French Canadians, Kentucky hill people, West Virginia miners. Whether it is happening in the same way for Negroes, Indians, or Mexican-Americans, however, is a question full of doubt and anger.

Unfortunately, our reliable knowledge of American migration breaks down at this point. Some analysts feel that every group of migrants to American cities (whether from overseas or inside the United States) goes through the same general process of economic and social integration into the city's life, and that differences in the migrants' wealth, culture, and acceptability to the host population simply accelerate or retard the general process. Others think that this was true of the great waves of migration from Europe five to ten decades ago, but that decreasing demands for unskilled labor and solidification of racial discrimination in our cities have changed the situation entirely. Still others argue that every national or racial group has worked out a somewhat different arrangement with American society, depending on the state of the economy at their arrival, on the enemies they faced, and on their resources, culture, and leadership.

We can see some of the reasons for this disagreement by examining three prominent recent groups of migrants to American cities whose members have often arrived with little money and few of the skills which urban life rewards. Let us look at Puerto Ricans and Appalachians briefly, and at Negroes in greater detail.

Since Congress radically restricted overseas immigration in the 1920's, few people without high educational or occupational qualifications have been able to enter the United States permanently. Puerto Ricans are an exception, because their land's status—first as a colony, then as a more autonomous Territory, and now a Commonwealth—gave them American citizenship and the right to move freely to the mainland. Few came until the end of World War II. Then the opening of inexpensive air transportation between San Juan and New York combined with a strong demand for workers in New York to attract thousands of migrants per year. The new movements from Puerto Rico to the United States ran around thirty or forty thousand per year until the 1960's, when they dropped to less than half that rate. The actual number of people migrating was much larger, since families and individuals moved back and forth easily and constantly. The number of people going in one direction or the other depended heavily on the relative prosperity of Puerto Rico and the United States at the time; things were good enough in Puerto Rico in 1963, for example, to produce a net movement of over five thousand persons back to the island.

By that time, not only New York, but also Philadelphia, Boston, and other East Coast cities had significant Puerto Rican settlements, complete with shops, churches, and Spanish-language movie houses. Puerto Ricans unquestionably started at the bottom in most of those cities. They had the worst jobs, the highest unemployment rates, the lowest incomes. Their very high birth rates, furthermore, kept Puerto Rican families large, meant that a large proportion of their members were outside the labor market, and made them exceptionally dependent on public assistance to supplement their incomes. Despite slow shifts toward prosperity, these things are still true.

Yet the coming of the Puerto Ricans is too recent for us to assume that they are stuck at the bottom. Many of the traditional early signs of a group's success in America are already appearing among the Puerto Ricans of New York: Puerto Rican baseball stars, politicians, and teachers; prosperous small businessmen; people with good jobs leaving Spanish Harlem behind and moving into richer neighborhoods of no particular nationality. So far it is still possible (if by no means proved) that Puerto Ricans are following more or less the same path as their predecessors.

Appalachia is a large band of mountains, hills, and adjacent river plains which cuts diagonally across the eastern United States. It covers important and backward parts of the states between North and South—Kentucky, Tennessee, and others. Pioneers pushing west from the Atlantic coast settled the region a century and more ago. They and their descendants organized it in small family farms growing enough to keep their residents alive, and not much more. Early in this century coal mines opened up through much of the region; they provided jobs for many of the extra hands the prolific families produced. They could not absorb all the excess population, however. The region has been exporting migrants to northern cities like Cincinnati, Detroit, and Chicago for decades.

Since World War II, however, the automation of some mines, the closing of many others, the mechanization and commercialization of farming, and the growth of modern industry along rivers like the Ohio have driven and drawn families by the thousands from Appalachia. Like the Negroes who came from the plantation areas farther south, these white migrants had little education, few job skills of particular use in the big city, and not much experience with urban life.

Most of the people leaving Appalachia went where they already

had close friends and relatives. The friends and relatives re-
cruited them, just as the newcomers in turn would later recruit
more migrants from back home. From the middle 1950's, they
have formed their own villages within northern cities, notable for
their overcrowded, rundown houses, their country music, their
local mountaineer bars, and the used cars kept in condition
for the trip back home. For even more so than the Puerto Ricans,
the people from Appalachia often dream of returning home,
often make trips back, and often stay there for good. As a result,
they are one of the most mobile groups in American cities. In
Chicago, which may have 30,000 recent migrants from Appalachia, one school in an Appalachian neighborhood had 1,500
children who entered or left during the school year for every
1,000 who stayed the entire year.

Because they are transient, because the people from the same
region who make good in the city aren't labelled as "hillbillies,"
and because the migration is so recent, it is hard to say what
course the integration of people from Appalachia into American
urban life is following. Careful follow-up studies of earlier migrants from rural Kentucky indicated that after some trial and
error most of them settled down, remained in the city, acquired
better jobs than they had at home, and began to disappear into
the general population. Whether in a time of accelerating flight
from Appalachia and continuing expansion of Appalachian
communities within big cities this will keep on happening is still
an open question.

The twenty-two million Negro Americans include a vastly
larger number of urban migrants than the Puerto Ricans or
the Appalachians. Their fate worries many Americans, white
and black. Many of the worries are confused by the false notion
that Negroes are like the recent Appalachian migrants in coming
mainly from southern farms. From World War I on, a huge
number of Negroes *did* move from the rural South to big cities
of North and South. But the big move is almost over. Within the
United States, Negroes are now more heavily concentrated in
cities than whites are. They are especially concentrated in big
cities like Chicago, New York, and Los Angeles. Like so many
popular ideas about cities, the notion of the Negro as an

urban neophyte comes from a confused memory of events already past.

While the great move out of the South was going on, it formed three great streams: the first up the East Coast through big cities like Washington, Baltimore, Philadelphia, and New York; the second from the Gulf of Mexico toward the Great Lakes and other big cities such as St. Louis, Cleveland, and Chicago; the third westerly from Texas and Oklahoma to California. Nowadays, however, more and more Negro migrants are moving from one northern metropolitan area to another; fewer and fewer are moving directly from the rural South to cities of the North and West. Malcolm X came to Harlem like many other Negro youngsters, but he came from near Detroit via Boston. During the late 1950's the majority of Negro migrants to big northern metropolitan areas like New York or Boston were coming from other metropolitan areas. Only a fifth of the 1960 nonwhite population of American metropolitan areas consisted of persons born on farms: even fewer of the people now on the move come from farms.

If the persistent vision of Negroes as displaced sharecroppers didn't get in the way, this wouldn't really be hard to understand. For one thing, not many Negroes are left on the farm. In 1960, there were little more than 200,000 Negro farmers in the states of the Deep South. Just over half of them were sharecroppers. Over a third of those farmers left the land during the following five years, and took their families with them. But as much as that massive exodus changed the character of the Southern rural population, it did not overwhelm the big cities to which the departing farmers were moving. Even if every single Negro living on an American farm in 1960 had moved to a metropolitan area within the following five years while other kinds of migration kept their pace, Negroes coming from farms would still have been a minority among the Negro migrants into the average metropolitan area.

Besides, job opportunities attract many more people from one metropolitan area to another each year than they induce to move into metropolitan life for the first time. Metropolitan residents have more of the skills and information which make long-distance migration feasible. They are more often involved in national, rather than local, labor markets. Since highly skilled occupations tend to produce geographically extensive labor markets, as the average occupational level of Negroes has risen in recent years, so has the frequency of their long-distance city-to-city movement in response to better job opportunities.

What about the differences between Negroes and whites in these respects? Even today, Negro migrants come from farms and from regions (like the Southeast) with generally low educational levels more often than white migrants do. These geographical differences are disappearing fast. Almost everywhere in the United States, Negroes, on the average, get less education and hold poorer jobs than whites. These differences are far from disappearing.

No one should be surprised, then, to learn that the average Negro migrant comes to the city with less education and occupational skill than either the white migrant or the bulk of the urban population. But compared to the Negro population already in the city, the average Negro migrant has a distinct advantage in age, occupation, and education. In short, once we make allowances for the national pattern of discrimination, we find that the same general rules hold for both white and Negro migration.

Now, this is a hard conclusion for many Americans to swallow, because they have the habit of blaming so many of the Negro American's troubles on migration. For example, many people have noticed and deplored the high rates of desertion, separation, divorce, and illegitimacy among Negro families. One widely-accepted explanation is that the strains of migration to big cities from the South commonly broke up families. But there is really no general evidence that long-distance migrants in America or anywhere else have more unstable family lives than other people. Such wisps of information as we have about recent Negro migrants to big cities suggest that their families may actually be more stable than the rest. The instability of Negro families is more likely created by the grinding unemployment and economic insecurity they face in the city than by the disruptive effects of migration.

Likewise, many people have explained the fact that American

Negroes have higher rates of conviction for major crimes than whites do by pointing to the personal upsets and deprivations produced by migration. Here the evidence is a little stronger than for family stability, and again it points the other way: recent migrants are less likely to commit crimes than are long-time residents of the city.

Finally, the great riots in the Negro ghettos of Los Angeles, New York, Detroit, and other big cities since 1963 have stunned and puzzled many Americans. Observers, both black and white, have often felt that one of the main factors behind the riots was the venting of frustration by disappointed newcomers from the South. The information from the Los Angeles riot of 1965, however, showed that the rioters arrested were mostly long-time residents of the city. Furthermore, very few of them had ever been in trouble with the law before.

The picture which is taking shape, then, does not show displaced individuals and families shattered by migration and therefore taking up crime and violent protest. It shows almost the opposite. It seems to take Negro migrants quite a while before they adopt the ways of the city; that means crime for a few, unstable family lives for more, participation in racial protest for others, wealth, power, and success for a handful, quiet misery for a great many. Migration does not produce these effects. Organized discrimination does.

The differences in opportunity between Negroes and whites appear from the moment of arrival in a new city. The white migrants to an American metropolitan area settle all over it, and especially in the suburbs. The Negroes go overwhelmingly to the central sections of the central city. Why? Well, the multiple forms of discrimination which make it so hard for Negro families to rent or buy housing in predominantly white areas produce much of the difference. For the Negro newcomer to a city, the main choices are the ghetto, the area next to the ghetto, or a fight. Other factors reinforce this segregation. By and large, housing in American metropolitan areas becomes more expensive with increasing distance from the center; people with little money to spend therefore cluster near the center. With average incomes barely half those of whites, Negro families rarely can afford to go very far out from the center. The problem of traveling to work in the central city enterprises which employ Negroes in any number sets another limit on the choice of dwellings. Finally, the tendency of those migrants who come to the city through contacts with friends or kinsmen to settle first with them or near them, as well as the tendency of other Negro families to seek protection and familiar surroundings near the ghetto, add a measure of self-segregation.

Few things are now working against this formidable phalanx of factors promoting segregation; some prosperous Negro families do move out of the ghetto, the legal supports of housing discrimination are falling away, and in the long run the fact that on the whole Negroes share the standard American preference for solid individual houses and gardens will no doubt win out. But up to now, American metropolitan areas have shown no real signs of desegregating. Migration and segregation have produced two fundamental changes in the character of American metropolitan areas, one unmistakable and the other easy to miss. The unmistakable change is the rapid rise in the proportion of the central city population which is Negro. If recent rates of increase continued until 1980, Chicago's population would be almost half Negro, Cleveland's 55 per cent Negro, and nineteen out of every twenty persons in Washington would be Negroes.

The subtle change in the character of American metropolitan areas is the increasing whiteness of the suburbs. Huge numbers of whites have moved into metropolitan suburbs, but almost no Negroes. Thus the share of Negroes in the suburban population has fallen. Increasingly white suburbs ring increasingly black central cities. The pattern of migration to American cities has obviously helped create the drastic and hurtful segregation of white from black. But without extensive discrimination in employment, education, and housing the process of migration would not operate as it does. The comparison among Puerto Ricans, Appalachians, and Negroes leaves unclear whether we ought to expect Negroes to follow the paths out of central city ghettos taken by Italians, Poles, and Irishmen. Still, the comparison makes two facts all the clearer:

1. The heavy presence of racial discrimination makes the exit from the ghetto slow and painful.

2. Many of the circumstances thought to be consequences of migration are really effects of the organization of life in the city itself.

It is possible, although by no means certain, that over the next few decades united Negro action and determined governmental intervention will wreck the structure of discrimination in jobs, schools, and housing. If that happens, much of the racial distinction, and racial tension, between cities and their suburbs will disappear. Even after that, however, the general distribution of jobs, housing, and land uses within American metropolitan areas will tend to concentrate the poorest and most alien newcomers downtown and the wealthier ones outside the center.

At the same time, we can expect the migrant from the farm —and even from the small town—to virtually disappear. For a great many reasons, it is unlikely that the rural migrant from overseas will replace him. A constantly mounting proportion of the new arrivals will be lifelong urbanites from other American metropolitan areas. An increasing number will come from metropolises outside the United States. The distances most of them travel from the old city to the new will increase; moves from coast to coast will become commonplace.

Somewhere in the future there is a possible alternative to the ceaseless flow of persons from city to city. Modern means of communication are making it increasingly practical to transmit only part of a man—his words, his voice, or his image—over long distances. So far the availability of substitutes for transmitting the whole man, like television and interlocking computers, has neither slowed down American internal migration nor stayed the rise of travel by automobile and aircraft. The substitutes could well become so workable that it would no longer make sense for men to travel three thousand miles for a meeting, transfer from one office to another, leave home to go to work, or even migrate in order to take a new job. An organization with its communication center in Chicago might have most of its workers on the Atlantic and Pacific coasts. By that time, we could reasonably expect the long-distance migration which is so much part of the present American scene to subside, and the sheer attractiveness of a particular city to become a much larger determinant of migration to it. Over a much longer run we could reasonably expect the same pattern to cover the entire world.

# 14 RACIAL SEGREGATION AND NEGRO EDUCATION

## Thomas F. Pettigrew

Two of the most pressing domestic problems facing the United States today—the deterioration of both its central cities and its race relations—often converge. In no realm is this convergence more evident than in the segregated and inferior education of Negro American children—the topic of this discussion. The painful truth is that the United States has failed to adequately educate fourteen generations of Negro Americans and is currently in critical danger of failing the fifteenth generation. But the character of this failure has radically changed in recent years, and it is now deeply enmeshed in the problems of our cities discussed throughout this book.

## PAST FAILURES OF NEGRO AMERICAN EDUCATION

The first ten generations of Negroes in America had slavery for a teacher. Save for rare exceptions, only the minority of Negroes who were free had any access to education prior to the Civil War. Indeed, many states of the American South had explicit laws forbidding the formal instruction of slaves, for it was realistically feared that the tutored Negro was a potential leader of revolts. The end of slavery and the initiation of a so-called "Reconstruction" era from 1867 to 1876 saw an earnest attempt to bring formal education to Negroes. Idealistic white Northerners, often paternalistic, perhaps, but motivated by a type of Peace Corps spirit of the period, opened up schools for Negroes throughout the South. The response was overwhelmingly enthusiastic. Ex-slaves of all ages flocked to these rudimentary institutions to learn to read and write. Having been long denied, education assumed a special appeal worth great sacrifice to acquire. And this strong desire for formal training remains to this day a general, if often overlooked, characteristic of Negro Americans.

Higher education for Negroes also traces directly back to this era. Today there are approximately 120 predominantly Negro colleges and universities in the United States, almost all of which are in the South and most of which began during Reconstruction. Roughly three out of every four Negro Americans with some college training today received it in one of these institutions, and about three out of five Negro Americans now in college are enrolled in them. With a few notable exceptions—such as the Atlanta University complex in Atlanta, Georgia, Fisk University in Nashville, Tennessee, and federally-supported Howard University in Washington, D.C.—these colleges are small, ill-equipped, and poorly financed. Usually they boast an interracial faculty but a virtually all-Negro student body. Yet they have for a century offered to many Negro youth their only hope for advanced study; and they have served as the chief training ground for Negro leadership and the significant and growing Negro middle class.

Public education for white as well as Negro Southerners did not take root until after the Civil War. As the poorest region of the nation, the South found it especially difficult to finance such a broad service to its people; and the rural South, where Negroes were still concentrated, found it almost impossible. Complicating matters further, rigid adherence to racially separate schools meant that the meager regional resources for public instruction were disproportionately employed for white education. Too poor actually to develop one school system of quality, the South had no means whatsoever of developing two racially separate systems of quality.

Once the reforming fervor of Reconstruction ended in 1876, the North eased its pressure upon the South for racial change. Reconstruction was in many ways unsuccessful, because, in ignoring the economic needs of the defeated South and its newly freed Negro citizens, the era's efforts did not "reconstruct." An early version of a "Marshall Plan" was required, but nothing like it was attempted. Consequently, conservative whites regained power by the late nineteenth century, and their discriminatory racial policies were allowed by the nation. In education, this meant racially separate schools, or even no schools at all for many rural Negro youngsters.

In the 1890's, the Supreme Court of the United States even lent legal support to racially segregated public facilities. In a famous 1896 case involving railroad car segregation in Louisiana, *Plessy v. Ferguson,* the Supreme Court made possible the doctrine of "separate but equal" racial facilities. This doctrine furnished the rationale for southern segregation of and discrimination against Negroes for fifty-eight long years. In fact, the doctrine was a notorious legal fiction: public facilities in the South remained distinctly separate and most *un*equal. Southern state laws required segregated schools during these years; thus, the failure to adequately educate Negro Southerners in the first half of this century derives from a legal fiction and so-called *de jure* segregation.

## THE CHANGING CHARACTER OF THE PROBLEM

As detailed in other chapters of this book, the Negro American has been rapidly becoming an urbanite throughout this century. While the inadequacy of public instruction for Negro children was largely a phenomenon of the rural South, it has now become a basically urban problem. America is still failing to educate fully its Negro citizens, but the scene and character of this failure has radically changed. While the vast majority of Negro Americans in 1910 were agricultural peasants in the rural South, the vast majority today are residents of cities— particularly the largest cities throughout the nation. Now there are more Negro citizens in metropolitan New York than in any southern state; more in metropolitan Chicago than in the entire state of Mississippi; and more in metropolitan Philadelphia than in the states of Arkansas and Kentucky combined.

Altering the nature of the problem, too, is the shift in the basis of racially separate schools. After whittling away at what "equal" actually meant in the doctrine of "separate but equal," the Supreme Court in 1954 made the final step and ruled that "separate educational facilities are inherently unequal." Sweeping and important as this ruling was, much of the world overinterpreted its immediate impact. Segregationists had many delaying tactics at their disposal, largely because of the traditionally strong state and local powers over public education in America's federal system of government; and segregationists have employed all of these tactics, including physical intimidation. Hence, progress against *de jure* school segregation—that formally required by southern state law—has been pitifully slow. But it has occurred bit by bit over the past decade and a half. While virtually no Negro children attended public school with white children in the former Confederate South prior to 1954, about one in six does today.

Yet racially separate schools throughout the country are a more serious problem now than even in 1954, at the time of the Supreme Court's historic ruling. For while the South has made limited progress in eradicating *de jure* segregation, the nation has allowed a swift increase in so-called *de facto* school segregation. Unlike legally supported *de jure* segregation, *de facto* segregation is so titled because presumably it is the unintended result of natural shifts of population and residence. Actually, *de facto* separation is a misnomer, since it is often the intended result of local governmental action. It differs in practice from the South's old system of *de jure* segregation, then, largely by not being required by blatant state statutes.

To sum up, Negro American children are still being woefully shortchanged educationally; but the locale and basis for this situation have markedly shifted. Where first the causes were slavery and later *de jure* segregation largely in the poor rural South, it is now primarily a result of more subtle *de facto* segregation in the country's largest urban centers. To place this transformation of the issue loosely into an international context, one might say that America's racial problems have moved from resembling those of the naked oppression of the Republic of South Africa's system of apartheid toward those of England's more subtle and urban-centered discrimination against its colored citizens.

## PRESENT FAILURES OF NEGRO AMERICAN EDUCATION

What, then, is the picture of Negro education in American cities at the present time? The basic fact has already been mentioned: Negro children are still largely found in segregated schools. Consider data from a 1965 national survey: Two-thirds of all Negro pupils in the first grade of the nation's public schools and half of all Negro pupils in the twelfth or final grade were enrolled in schools with 90 to 100 per cent Negro student bodies. Moreover, seven out of eight of all Negro students in the first grade of public schools and two-thirds of all Negro students in the twelfth grade were enrolled in predominantly Negro schools. Though different in magnitude, the regional discrepancies no longer alter this picture significantly: while 97 per cent of Negro first-graders in public schools of the urban South attended predominantly Negro schools in 1965, the figure for the urban North was a high 72 per cent. In turn, white American children are even more segregated from Negro American children: 80 per cent of white public school children in both the first and twelfth grades were located in 90 to 100 per cent white schools in 1965.

Moreover, the separation is increasing. In Cincinnati, Ohio, for example, seven out of every ten Negro elementary children in 1950 attended predominantly Negro schools, but by 1965 nine out of ten did so. And while Negro elementary school enrollment had doubled over these fifteen years in Cincinnati, the number in predominantly Negro schools had tripled. This pattern of growing separation is typical of American central cities, the very cities where Negro Americans are concentrated in greatest numbers.

There are at least three major causes for this increasing degree of *de facto* school segregation: (1) the fragmented, antimetropolitan nature of school district organization in the

United States; (2) the effects on the public schools of private and church-related schools; and (3) intentional segregation by design of local authorities, as discussed earlier. The first of these factors becomes apparent as soon as we compare public school organization and current racial demographic trends. Incredibly enough, there are 27,000 completely separate school districts in the United States. There are, for instance, over seventy-five school districts in the Boston, Massachusetts, metropolitan area alone and ninety-six in the Detroit, Michigan, metropolitan area. Add to this the fact that over 80 per cent of all Negro Americans who live in metropolitan areas reside in the large central cities while slightly over half of all white Americans who live in metropolitan areas reside in suburbs, and the racial separation by district becomes apparent. Racial housing trends are not encouraging and offer no hope for relief of educational separation in the next generation. Consequently, America would face an enormous problem of *de facto* school segregation even if there were no patterns of intradistrict separation by race.

But the nation also faces the task of overcoming sharp racial segregation within school districts. In cities with large Roman Catholic populations this intradistrict segregation is unwittingly increased by the absorption of many white children into the Church-run system. Since only about 7 per cent of Negro Americans are Roman Catholics, a large Church school system necessarily limits the available pool of school age white children for a central city public school system.

Finally, southern and northern segregationists in American political life make the problem of intradistrict segregation worse by openly advocating separation, careful placement of school attendance areas so as to maximize *de facto* school segregation, and resistance to measures which would at least begin to alleviate the problem. But while the public resistance of anti-Negro politicians gains the newspaper headlines, the most important reasons underlying the growing pattern of separate schools include the nation's inefficient organization of school districts and the effects of overwhelmingly white Church schools.

But how damaging to Negro education is school segregation?

Is it not largely a matter that mostly Negro schools have traditionally had poorer facilities? These questions are often raised in America these days as the problem of surmounting the growing segregation of the nation's schools becomes increasingly more difficult. Consequently, the United States Commission on Civil Rights conducted a thorough research investigation of the question of the effects of educational segregation by race in two realms: academic achievement of Negro children, and the racial attitudes and preferences of both Negro and white children.

Apart from other factors, the Commission data strongly suggest that the academic performance of Negro children in predominantly white classrooms is superior to that of comparable Negro children in predominantly Negro classrooms. Moreover, the achievement of white children in interracial, but still mostly white, classrooms is at least as high as that of comparable white children in all-white classrooms. In academic terms, then, Negro children gain while white children do not lose in the predominantly white, interracial classroom. In racial attitudes, both races benefit from desegregation. In contrast with comparable students in all-Negro or all-white schools, both Negro and white children are less racially prejudiced and more frequently prefer interracial settings when they have experienced desegregation. Understandably, these attitude effects are most notable among those with close friends of the other race. And this raises an interesting distinction between merely desegregated schools and integrated ones. A desegregated school has an interracial mix of students, usually with a majority of whites. An integrated school, however, is desegregated but in addition boasts genuine cross-racial acceptance and friendship within its faculty and student body. Desegregation, therefore, is only a prerequisite for the real goal: racial integration.

Both the academic and attitudinal benefits of integrated education are strongest, the Commission on Civil Rights found, when the interracial experience begins for the Negro and white child in the early grades. The practical difficulty raised by this important finding is that it is the elementary schools that are the most segregated in American cities, since their smaller attendance areas are the most closely linked to racially segregated residential patterns. Yet any true solution for this problem must include the beginning grades if it is to maximize the benefits of interracial schooling.

## IDEAS AND PROSPECTS FOR SOLUTION

Remedies for the segregation of urban schools follow directly from the causes of the problem discussed previously. To achieve integrated education throughout America's metropolitan areas, the nation will have to act in metropolitan terms. This metropolitan approach to the problem need not entail a single, enormous metropolitan school district, but it will require far more metropolitan cooperation in education than is typically the case in the United States today. Federal and state support that is specifically designed to encourage and reward metropolitan cooperation is urgently required. The nation's central cities cannot solve their racial problems alone, even with huge infusions of federal money. And suburbs cannot remain the proverbial "white noose" around the central city Negro's neck if the country's racial problems, educational and otherwise, are ever to be dealt with realistically. Once the issue is put into a metropolitan context, incidentally, the population patterns are more encouraging for the possibility of urban integration: almost the same proportion of Negroes as whites —roughly seven out of ten—reside in metropolitan areas, which means that Negroes are no more represented in America's 212 metropolitan centers than their one-ninth proportion nationally.

In addition, more cooperation between public and private school systems is called for, a trend already underway that also needs financial encouragement from federal and state programs. Both systems stand to gain immeasurably from such cooperation. The involvement of Roman Catholic youngsters from Church-operated schools in regular public school programs could save the Church considerable money, while it would also provide white children to heavily Negro central city systems for greater desegregation.

What design do these new structural arrangements for American public education suggest? To make widespread desegregation possible, future urban schools would have to be metropolitan in character, involve public and private school cooperation, draw their students from broad attendance areas, and introduce a marked upgrading of educational standards and facilities together with desegregation. My own favorite possibility for designs meeting these criteria is best described as "the metropolitan educational park." Large educational campuses, strategically located on major transportation spokes for both suburban and central city attendance, could include the full range of school grades and through efficiency of scale provide more and better facilities than now dreamed of in American public schools. Accommodating 12,000 or more children of all ages, such an educational complex might well contain two or more high schools for students from fifteen to eighteen years of age, five or more junior highs for students from eleven to fourteen, and twelve or more elementary schools for students from five to ten in a university campus-like setting on a tract of one hundred acres or more.

All four of the criteria for aiding desegregation are met by such a plan. The well-designed educational park could make both metropolitan and public-private school cooperation more feasible, draw its students from a wide geographical area in the shape of a pie slice covering a range of races and social classes, and make possible a marked upgrading of public education. Yet the racial gains of such educational parks would be only part of their numerous educational advantages. The centralization of resources together with both construction and operating cost savings permit a fresh new start in public education. Exciting innovation in both the physical and social structures of the schools would be more possible. The economy of scale would also allow wider course offerings and special facilities and programs not now possible. And centralization would attract other public services to the site—such as public libraries and public health offices—and make coordination with higher education more attractive.

As with any plan, there are also potentially serious disadvantages with such a sweeping reorganization of urban education. First, the size of the complex raises the specter of massive and frightening impersonalization—what might be termed "the Kafka problem." Certainly a poorly designed metropolitan educational park could raise such issues, as they have already been raised by ill-planned and enormous American public schools in existence. But a well-planned complex with multiple individualized school units smaller than the average school today could actually turn back the clock to a time of less educational impersonalization. A second possible disadvantage is the loss of parental involvement that is presumed to be strongest in local schools. While it is questionable how many local schools in America actually command high parental interest, there is no reason why the carefully planned park could not also attract parental involvement—especially once it is demonstrated that it is providing the children with a better education and that it offers facilities so superior as to invoke adult interest and use.

Why, then, are there not such educational parks throughout the United States now? The reason is simple: the capital outlay is typically too expensive for local school districts to finance. Massive federal support for public school construction is necessary before they can be built. The political climate for such federal aid is now favorable; but Vietnam war costs make such aid impossible at the present time. Like solutions to many pressing domestic problems facing the United States today, metropolitan educational parks will not become a reality until the Vietnam war mercifully ends. In the meantime, *de facto* racial segregation of the nation's urban schools will steadily grow more severe.

## JUST AN AMERICAN PROBLEM?

Let me conclude by considering the wider implications of our discussion. Is segregated schooling just an American problem? I think not. In its tragic racial aspects, of course, the American situation is unique. But, in the increase of intergroup tension through separation, this problem is by no means unique. Tensions between members of different social classes, religions, tribes, and nationalities unfortunately abound throughout the world. And a universal feature of all of these situations is that separation—whether *de jure* or *de facto*—in time exacerbates these intergroup tensions.

One of the most effective societal institutions for breaking down separation and hostility between two groups is the school. Free of the bitterness of the past, young children are best able to sense the basic humanity of each other regardless of group identification. Furthermore, societies around the globe otherwise perpetuate through segregated schools inequities of access to quality education.

Several years ago I attended an international conference of social scientists. And I applied this reasoning to argue for school integration in India between castes and language groups, in Japan between Koreans and native Japanese, in the Philippines between Christian and Muslim, in New Zealand between Maori and European, in all countries between diverse social classes. Other conferees were quick to agree that racial integration in the United States was indeed a necessity, but equally eager to insist that their national situation was unique and school integration for them unwise. They only succeeded in convincing me that it is far easier to advocate sweeping reform in another nation than to accept it in your own. The United States is in desperate need of racial and social class integration of its urban schools. And I suspect, if you think about it honestly, school integration between diverse groups is needed in other countries, too.

# 15 THE CHALLENGE OF URBAN EDUCATION IN THE UNITED STATES

**Robert A. Dentler**

Today, more than thirty million people live under urban conditions in a continuous area that stretches more than six hundred miles down the Eastern seaboard, from Boston to Norfolk. By urban conditions, we mean that these people are culturally very diverse, yet very interdependent economically and socially. The same vast number of persons could sustain human life within the same geographic area, though at a very reduced level of subsistence, under rural conditions. It is not numbers, therefore, that determine their urban life situation.

Similar but smaller, less densely settled metropolitan areas dot the entire North American continent. Twenty such areas are dominated by great central cities that house more than half a million residents each. Another hundred areas contain cities of more than one hundred thousand residents. Each metropolitan area also includes dozens of satellite cities, suburbs, and semiurban fringe developments.

American urban achievements and failures are as old as our oldest cities. From the early eighteenth century through the close of the nineteenth, American cities served as incubators of native culture and creativity; as importers of the high culture from Europe and the Orient; as vital stimulators of technological and commercial development. Through the same period, American cities were congested, polluted in air and water, ridden with crime and disease, and stupified by political chaos and corruption. If one uses an absolute standard, the growth of urban culture on the one hand and the application of solutions to problems on the other have undergone little change since 1900. If a relative and humanistic standard is applied, then the contemporary urban situation is normal, not pathological; promising and emergent, not morally deplorable.

The vast urban region from Boston to Norfolk is a prototype of the emerging state of postindustrial society throughout the North American continent and elsewhere. The units of the prototype are high population density, with shifting distribution across the region and out of the central cities, ever more elaborate economic and political structures, and social cohesion through mass culture.

Like other urban institutions, schools evolved during a period of comparatively low density, comparatively simple conditions of a factory economy and limited local government, and a social cohesion maintained through mechanisms of social class, ethnic group membership, and corresponding distinctions in styles of living. In the late nineteenth century and well into the present period, city schools in New York, Buffalo, and Pittsburgh, for example, equipped the children of European immigrants for livelihoods in the factory and supporting business institutions.

Whether a child remained in school or withdrew, what is more, made only a small difference in his life chances, for the city-based industries of that day required a cheap and abundant labor force located close at hand. As late as 1920, 80 per cent of American youth left school by age 13 with no more than an eighth grade education. For its time, however, this education was good enough to provide literacy and the ability to handle rudimentary arithmetic.

The great central cities gradually evolved impressive high schools and colleges to supplement this basic system. Gifted students, regardless of social class, received increasingly sound advanced instruction, and technically-oriented youth began to benefit from the growth of technical and vocational schools tailored to the production of white-collar workers, factory foremen, draftsmen, and craftsmen. The greatness of this system resided in the fact that it was free and universal. Wherever a newly arrived family settled, one of the first institutions its children faced in the cities was the local school. For the diligent student, the local school offered Americanization and employability. For the gifted, diligent student, the same system offered cultural assimilation and swift upward mobility.

No public urban institution in America worked very well before or during the early part of the twentieth century. These were years of extremely rapid and often exploitative economic development. But public belief in formal education as an instrument of individual opportunity was strong and, except during periods of financial recession, investments in school plant and teaching staffs were high. Urban public education modestly yet steadily increased its ability to serve urban children and youth. Between 1920 and 1960, for example, the proportion of urban youth completing twelve grades of school increased from 20 to 60 per cent; and 40 per cent were continuing on into college.

In spite of steady gains, however, urban schools at all levels today carry out their tasks in a challenging, frustrating milieu of change. For example, whenever the lower school staffs manage to get attuned to the cultural styles of their students, their student bodies change in response to changing patterns of residence. Schools in Harlem, for example, reached something of a racial balance in the years from 1920 through 1929. During the latter part of this period, Harlem became a black ghetto, and in the 1930's it became the extreme victim of the Great Depression.

White teachers, only recently adjusted to teaching the children of German and Russian Jews and Italian immigrants, struggled to adapt increasingly scarce resources to the new task. But their techniques and their cultural assumptions, even if judged against a relative standard, were poorly fitted to the change. Today, thirty years later, Americans are acutely aware of the scope of failure in educational services in Harlem and similar ghetto communities throughout the urban North. Even the small improvements made by valiant educators during three decades seem slight when measured against the ever more pressing demands for educational progress.

In balance, there seem to be thirty years of mutual failure

in the schools of Harlem; mutual failure, for many teachers are unable to teach effectively and many children do not learn. This failure is not uniform, for Harlem and smaller ghettos from Boston to Norfolk do have some schools with gifted principals and effective teachers, men and women who persist successfully, who adjust and cope with change in pliable, imaginative ways in spite of everything.

Educational failure in urban lower-class neighborhoods is the most widely advertised feature of the urban school scene. It has received great public attention as an aspect of the civil rights revolution because Southern rural migrant families of Negro Americans who settle in the great cities of the North have demanded improvements in the quality of school services. And the older difficulties in fitting instruction to different family and pupil cultures have been intensified by racial isolation and rising expectations.

Educational failure in predominantly Negro city schools is further compounded by difficulties in tailoring instruction to the needs of other class and ethnic minorities. Hundreds of thousands of Spanish-speaking pupils—Mexicans, Cubans, and Puerto Ricans particularly—need bilingual instruction that is rarely provided. Second and third generation Polish-American and Italian-American students are often either isolated in overcrowded parochial schools or confronted with programs that are not adapted to their special cultural requirements.

The paradox is that urban Americans are extremely diverse culturally, yet urban educational programs are bureaucratically managed and uniform in operation in ways that work against diversified programming. Hundreds of new efforts, many of them federally funded, have begun during the last decade to remedy this situation. Some of them operate from the dubious assumption that cultural differences ought to be ironed out, as this was once a prime mission of urban public education. But many are exploring new ways of fitting the program to the special requirements of students as public clients.

Progress is especially difficult at this time because of changes in related facets of urban school life. One of these is the unionization of teachers. Until quite recently, teaching was a relatively depressed occupation in the United States. Though wages have improved and the training required for entry into teaching has become more demanding, the prestige and esteem accorded the urban teacher have remained slight. For one thing, the public school teacher in American cities was originally an underpaid high school graduate. Later, she became a bureaucratic functionary, a civil servant. For a long time now, upward mobility for an urban teacher has consisted of initial service in a low-income neighborhood followed by stages of migration toward high-income neighborhoods. As suburbanization has drawn the wealthier households into the suburbs, this movement has drawn thousands of the best qualified teachers out of the cities and into the suburbs.

In the last decade, urban teachers have organized collectively and with militancy in an effort to improve their own control over their occupational destinies. Few observers doubt that within another ten years city and suburban faculties will operate primarily as unionized associations of professionals, in the craft guild tradition.

One effect of this change is a fragmentation of authority in the design and management of formal education, making adjustment to new demands on the part of parents, students, and employers even more difficult to achieve. It is possible, however, that unionization *could* produce a revolution in the status and role of teachers, thus enhancing the professional competency of the institution as a whole.

Another change contributing to the difficulty of steady evolution of urban education in America is the transformation of the occupational structure. The cities and their schools were designed historically to facilitate a factory and trade economy. Cheap and abundant labor is no longer crucial to the economy. The urban blue-collar working class has shrunk. The present need is for highly skilled, hence highly educated, workers capable of continuous readjustment to the changing technology of a global industrial system.

Just at the point when urban educators devised a system capable of educating most youth through the twelfth grade, this second industrial revolution has made conventional secondary education obsolete. In California, where change has been most rapid in this respect, most pupils in the Los Angeles and San Francisco metropolitan areas continue their education through at least two to four years of college or technical school. But even in these advanced systems, the lower classes and the children from ethnic minorities have not been equally well served. Their attainments lag gravely behind the necessary standard.

Urban schools are unlikely to change at a faster pace than other institutions under any conditions. Indeed, they tend to change slower than those services which, like water and highways, depend mainly upon the pace of economic and technological development. If we add to this the facts of economic and racial isolation of the poor and the ethnically different, the struggle of teachers for a better place in the sun, and the impact of the second industrial revolution upon manpower requirements, we begin to grasp the challenge facing the urban schools.

An awareness of this challenge has grown rapidly since 1955. Urban educational budgets have more than doubled in the last fifteen to twenty years. Research and development efforts have been introduced at public expense in the quest for solution. Yet political leaders are limited in what they can achieve on behalf of reconstruction. For example, the legal basis for school desegregation has been firmly established in New York, New Jersey, California, and Massachusetts. State boards and officers there require the establishment of attendance areas for schools which will avoid racial segregation or eliminate existing segregation. Efforts at integration have been validated repeatedly by the highest state and federal courts.

Nevertheless, the public mandate continues to be a mandate for continued racial isolation in the schools, combined, when pressed, with token changes of a superficial kind. Public schools are also often separated in American cities from the power and authority of city hall, and this separation is reinforced by the historical strength of policies that separate church and state.

As a result, the challenge facing urban education is well understood, but public indifference combined with fragmented authority make reconstructive policies and practices difficult to introduce in quick order. When reconstruction does occur, it will include three essential ingredients. First, schools will be operated on a much wider geographic basis, preventing segregation and stimulating cooperative unification of authority. This will help to bring schools into accord with metropolitan changes in all other respects, where old boundaries have been broken and interdependence heightened. Second, teachers and citizens will have an increased voice in the determination of policy and programs. As authority is unified, its mandate must be shared locally with citizens, teachers, and students, if the vitality of this human service institution is to be maintained. Third, with unified authority and increased public participation, the program of instruction will be adapted to the diversity of cultural and manpower requirements of the times.

All three transformations are occurring within urban schools today. Urban universities are building extension campuses out into the suburban areas, and more rural schools are devising extension campuses in the city. The busing of pupils of all ages over long distances is becoming commonplace. At the same time that teachers are organizing, parents are demanding an increased voice in school affairs. Even the old wall separating parochial from public schools has been breached at points.

At the instructional level, there are exciting, positive transformations taking place. These include the application of technology to learning and to school administration; new degrees of cooperation between higher educational institutions; and a constant gain in cultural transmission through the benefits of modern transportation and radio, telephone, and television. Among the many forces working to equalize educational opportunities and upgrade them, perhaps the most promising is the gradual emergence of a national coalition pressing for a system of education with relatively uniform standards of excellence but with high tolerance for different approaches to teaching and learning.

None of these transformations will prove satisfactory, however, unless the pace of change is sharply increased. The most critical challenge facing urban educators is the change in what citizens insist upon and what teachers know is possible. The most sensational urban educational failure—the mass production of virtually illiterate youths in ghetto schools—can be overcome. Progress toward this end will continue to be made. The question is whether the progress will be fast enough to keep pace with what seems possible technically, professionally, and financially.

The challenge is thus one of incompatible rates of change. By the time urban schools—including colleges and universities have been raised to educational adequacy, will another generation of youth have slipped by? Medical science and its benefits have given most Americans somewhat insatiable aspirations. We want schools to do for learning what the modern hospital does for illness. This is now technically feasible. But for political and economic reasons which include the cost of warfare and defense, it remains unachieved.

# 16 THE URBAN CHURCH

Andrew M. Greeley

Modern man has increasingly become obsessed with the pursuit of community—a pursuit which in its youthful version has become explicit and even, in the new psychedelic community, bizarre. So pervasive is the quest for community that it is easy to forget that the quest is a relatively recent one. The grandparents or the great-grandparents of contemporary man did not have to seek for community. They had it. Indeed it must have seemed to the more creative and restless members of peasant society that there was rather too much community.

But the move from the farm to the industrial city has involved the loss of both the social support and control which the small agricultural village provided for its citizens. This loss is the price that modern man has paid for the vast economic gains that urban industrialism has made possible. The citizen of the contemporary city is not about to relinquish the abundance and the affluence that his migration has made possible. And yet the struggle to compensate for those community dimensions which were left behind in the agricultural society has never stopped. For the first time, it would seem, modern men, or at least their sons and daughters, are persuaded that it is possible to have one's cake and eat it too—to have both affluence and community.

But while the quest for community is now explicit, it is not something that was discovered after 1950. The small, informal face-to-face group has had a tremendous influence within the urban industrial society, whether it be on the factory assembly line, in making political or marketing decisions, in choosing one's friends, neighbors, marriage partners, or one's clients and purveyors of professional services, or even in structuring modern military establishments. Nevertheless, whether it be something as large as an ethnic group or as small as a street corner gang, the modern attempts to regain or to preserve community in a rationalized, formalized, bureaucratized society have yet to satisfy modern man. The quest for community has been so intense that it is relatively rare for anyone to wonder whether community of itself is necessarily a good thing.

But it ought to be obvious that some kinds of communities facilitate the growth of the individual personality and the welfare of society, while others have harmful effects either on the individual or on society, or on both. Perhaps the most serious problem is that the many communities to which a man may belong need not necessarily be in harmony with one another. We hardly need be told that there is frequently deep conflict between the personal commitments of the businessman for his job and for his family, even to the extent that the executive's commitment to his secretary can be an obstacle to or a substitute for his commitment to his wife. In the peasant society there was a basic harmony between the two or three communities to which a man belonged—family, village, and church. The national society probably did not matter very much, save in time of war when fields might be devastated or sons might be pressed into military service.

In the modern city, the harmony between the local community and larger communities is not so easily preserved. The most serious disharmony, the one with which we will be concerned here, is the conflict between the local community (or "the neighborhood," as it is frequently called in the United States), and the larger metropolis. Far from these two communities to which urban man belongs having harmonious goals, it often seems rather that one must assume opposition between the goals of the neighborhood and the metropolis. Contemporary man is much more likely to be emotionally committed to the welfare of the neighborhood than to the welfare of the metropolis. From one point of view, therefore, the most serious structural problem that the modern metropolis faces is the lack of the technical, social, and ethical tools for communication among the communities which compose it. Man can desperately seek community for himself at the neighborhood level and create communities which are extremely helpful in meeting many basic human needs at the grass roots. But by the very fact of creating such meaningful communities he can simultaneously frustrate other less articulate needs which can only be served at a higher structural level. In the long run, the very success of the neighborhood community in meeting the social and emotional needs of its members may be the tools of its own destruction. The inability of the neighborhood community to communicate with the other communities which constitute the metropolis is a disaster for the neighborhood itself.

Even though there can be no doubt of the importance of the neighborhood to anyone who has lived in a city with large ethnic populations, we know surprisingly little about the impact of neighborhoods on American personality and social structure. It cannot even be said with any degree of confidence that all American cities have neighborhoods in the social psychological sense; it seems very likely that some inhabited areas within the cities do not, in fact, emerge as neighborhoods, particularly among the lower socioeconomic levels of the population where the kinds of organizational skills necessary to create neighborhood communities may very well not exist.

Future research may provide some of the answers to these questions. But at least in some cities, intense loyalty to local geographic areas leaves no doubt that these areas have become very meaningful communities in the lives of their people. When people move from the core city to the suburban fringe they begin to speak of the "old neighborhood" and the "new neighborhood," the former being the place where they grew up and the latter being the place where they live and are raising their own children. (Sometimes such terminology develops a curious twist of language: recently I heard a young couple living in Hyde Park, located near the central part of Chicago, refer to Kenilworth, the most exclusive of the Chicago residential suburbs where they were raised, as "the old neighborhood"; one suspects that most citizens of Kenilworth would be profoundly shocked.)

188

These local geographic areas can generate, as we have noted, intense loyalty and serve many varied practical and emotional needs of their citizens. As strong, cohesive communities they leave little to be desired. But unfortunately, satisfying personal relationships not only do not lead to a concern for the larger metropolis, but very frequently impede such concern. In recent unpublished research, James Barry tested the effect of voluntary tutoring work in the inner city on the racial attitudes of middle-class white adolescents. He found no significant difference between adolescents who engaged in such tutoring work and those who did not, and thus was led to conclude that the tutoring experience of itself did not necessarily lead to any change in racial attitudes. But when those tutors with strong loyalties to the neighborhood in which they lived were compared to those who got relatively less satisfaction from neighborhood communities, it was discovered that the latter were likely to be rather strongly affected by the tutoring experience. Commitment to the neighborhood community was an obstacle to social enlightenment flowing from the tutoring experience. Barry argues that strong neighborhood community ties produce "urban villagers" who are more narrow in their perceptions precisely because of the intensity of their community experience.

These urban villagers find it difficult, if not impossible, to communicate with those who function beyond the neighborhood level or to imagine any problems or the possibility of meaningful community which transcends their own limited geographic area. They have little sense of, or skills in, metropolitan problems. The metropolis does not become a community of communities, but rather a no man's land in which various neighborhoods and other power institutions compete one with another, not so much for dominance as for protection of their own interests and rights. In cities where a political organization is powerful, it usually plays the role of honest broker (honesty, of course, being a relative term), balancing the needs, demands, and rights of the neighborhoods and other institutions in some sort of precarious harmony. But the political organization necessarily lives in a world where power is a basic reality, if not the ultimate reality. A powerful neighborhood or alliance of neighborhoods, a powerful university or labor union group—or even a newspaper—is frequently able to force its will on the political organization in the absence of cohesive and organized resistance. Leaders of the political organization may indeed hold out some symbolic goal concerning the development of the whole metropolis and the possibility of the good life for the metropolis and for all the participating communities and institutions. In practice, however, the political organization must be too deeply occupied with balances of power to pursue seriously the task of forming a metropolitan vision and a metropolitan ethic.

In this frame of reference it is possible to view the civil rights battles as an attempt (and a frequently unsuccessful attempt) at communication between the Negroes or rather, more specifically, the civil rights leadership—and the white neighborhoods. But neither the social organization of the metropolis nor its ethical skills have developed to such an extent that communication between the Negroes and the white neighborhood communities has much chance of being peaceful and constructive. The message the Negro protest marchers invading white neighborhoods bring is obscure and ambivalent, and the response of the white mobs is predictable and, given the present development of the metropolitan area, probably inevitable.

The metropolitan community, unlike the family or even the neighborhood, need not be a community of love. It must merely be a community of civility. People need not communicate with each other at the metropolitan level with affection, but they must at least speak to others civilly. There are, it would seem, four major obstacles to such civil communication.

First, techniques of transportation and communication are not yet sophisticated enough to facilitate civil discourse. In most American cities telephone systems are reasonably efficient, though in times of crisis often are not adequate to the intensity of communication that is required. (In many non-American cities, even telephonic communication is at best a hazardous method of discourse.) It is at least plausible to contend that the inadequacies and inefficiencies of metropolitan transportation, both public and private, produce a level of frustration which is injurious to stability; also the sheer physical difficulty of moving people and resources from one part of the city to another restricts the possibilities of discourse. (Again, non-American and non-Western cities are faced with far more serious problems of internal transportation than are American cities.) But it is a curious irony that men come together in cities to combine their skills and create the good life for themselves, and yet the very closeness of their concentration frequently prevents the flow of men and ideas and resources which make the good life possible. While American cities have not yet ground to a halt, it is not inconceivable that they might. But non-Western cities have far more acute problems with far less in the way of resources to cope with the problems. On a world-wide basis, one is almost tempted to say that man may not be capable of providing the sheer physical means of transportation necessary to make urban life possible.

Second, in all probability technologies and techniques exist which, given the money and the will, could solve the physical obstacles to civil discourse in a metropolis, but in the absence of social organizational skills that could implement civility on a metropolitan basis, it is absurd to expect that the technologies and techniques will ever be utilized. In non-Western cities such as Seoul or Manila, the level of social organization skills required for effective metropolitan living is so minimal that it is dubious whether the aggregations of people, buildings, and equipment labeled under the name of the city can, in fact, be considered a city in anything but an analogous sense. Even American cities have not produced the kinds of institutions and organizations in which metropolitan discourse can occur in a fashion any more sophisticated or more urbane than a balance of power, or what Norton Long in *Polity* has called "the ecology of games." American cities are not masses of atomized individuals or nuclear families; they are aggregations of independent and normally uncooperative power structures with shifting patterns of allegiances based on the self-interest of each power structure on a given community issue. Such a social order may be a more sophisticated jungle than can be found in many non-Western cities, but it is still a jungle.

191

Third, it is doubtful, however, that we could argue that man does not know the kinds of social organizations which could facilitate metropolitan discourse. Social, political, and administrative sciences have reached a sufficient level of sophistication where they can, at least in theory, make extraordinarily plausible suggestions about how the metropolis might be more civilly organized than it is presently; but in the very act of making such suggestions, the behavioral scientist realizes all too well how naive they appear in the face of the realities of urban living. The development of man's ethical sense has not kept pace with the development of metropolitan complexities. Again non-Western cities illustrate the problem more clearly than do American cities. In many such cultures politeness and courtesy are traditional societal norms, at least in those areas of behavior in which the cultural tradition is expert. In such new kinds of behavior as driving a car in crowded streets, however, traditional culture has no ethical norm to apply, and traffic becomes a complete chaos which the American traveler finds first amusing, then frustrating, and eventually terrifying. In the United States we may be better, though only somewhat better, at handling the complexities of rush-hour traffic, but few of us are capable of conceiving of something as large as the Chicago and New York metropolitan areas as communities, much less imagining that there are ethical principles which might regulate the behavior of the citizens of these communities in the promotion of the commonwealth of the metropolis. The common man is not the only one lacking in the metropolitan ethic. Even the professional is overwhelmed by the immensity of metropolitan problems and is tempted to escape by proclaiming that a metropolitan ethic is impossible— a loss of nerve which could well indicate the death throes of urbanity.

Fourth, in the final analysis, however, the problem of establishing harmony between the local community (neighborhoods and neighborhood-like groups) and the metropolis is philosophical and theological rather than ethical; and modern man has yet to see a clear and compelling vision of the meaning of metropolitan civility or of the conditions of the good life in the metropolis. It has not been made clear to him that without the most complex kinds of cooperation and love, the metropolis cannot live up to its promise of providing greater opportunity for human fulfillment than man had previously dreamed possible. No one has yet phrased in cogent enough terms for modern man the argument that love and civility are the only way to prevent the metropolis from destroying both the vision which brought it into being and the men who have created it. Far from making the good life possible, the metropolis could easily produce a style of life which, while more physically comfortable, would be less human than what man knew in the peasant communes. Such a disaster has not yet happened, yet one is forced to conclude that we are far down the road toward that disaster, and that unless there is an emergence of a new urbanity, disaster is inescapable. Further, if such urbanity does not emerge in the Western world, it certainly will not have a chance in the non-Western world.

Even though it is frequently contended that he has become completely secular, modern man will probably turn at least once more to his churches as the only institutions capable of creating a new urbanity, of providing a vision of the possibility of metropolitan community. But the churches are caught in a desperate paradox. They have proven themselves reasonably adept at creating community at the grass-roots level; many parishes and neighborhood congregations have become very strong communities indeed. But the urban church has thus far been considerably less than successful in even comprehending the existence of the larger problems which transcend the local congregation. It has not produced the poets, the prophets, and the visionaries that the metropolis must have. On the contrary, it is hampered both theoretically and practically by the parochial structure—theoretically because a strong parish community is frequently no more capable of cooperating with anyone beyond its limits than is the neighborhood community in which its roots are sunk.

The church is therefore charged with devising a vision of metropolitan urbanity (as is presumably the university, which unfortunately long ago abdicated most of its meaning-bestowing functions). Such a challenge presents a fabulous opportunity for the church to prove that it still has some relevance to the modern world—a relevance which would ultimately be far more important than scurrying about on picket lines. But unless the church can find some way to transcend its parochially-oriented structure, it can hardly be expected to raise anything like a prophetic voice to sound out over the rooftops of the city.

If there is any small hope which might encourage us to believe that the metropolitan community might emerge, it is the appearance of the "expressway generation." At least some young people, most of them in their early and middle twenties, seem to be capable of "thinking metropolitan," even though they do so unself-consciously and inarticulately. The mobility that the automobile, the expressways, the jet planes, and the colleges and universities have put in their lives makes it possible for these young people not only to think of themselves as citizens of a city, but to behave in fact as though the whole metropolis was their "neighborhood." There is but impressionistic evidence for the appearance of this "expressway generation," a contemporary of and, perhaps, the successor to the "New Breed." From the "expressway generation" the new prophets will come, if they are to come at all. The churches will have to listen to their voices before it is too late.

# 17 POVERTY IN THE UNITED STATES

## Lee Rainwater

From the perspective of the material standard of the world at large, American poverty is a curious phenomenon indeed. Americans call a family of husband, wife, and two children poor if it receives about $3,000 a year or less in income, but that is an amount that many families in Western Europe would consider quite adequate, and an amount that many families in less industrialized countries would consider quite luxurious. A family considered poor in the United States, were it to transport its material possessions and household to many other countries in the world, would discover itself in terms of material affluence suddenly to be no longer poor, but well above average in economic standing. If a comparison is made on the historical rather than geographical basis, the same kind of seeming paradox obtains. Forty years ago well over half of American families received what today is considered a poverty income. Yet at that time few of the families involved would in fact have considered themselves poor, and poverty as a national problem was not at all such a central focus of public concern and policy as it is today.

Over the past seven or eight years the poverty problem has been of rising interest and common concern to American scholars and intellectuals, and from that interest and concern there has gradually grown an ever greater commitment on the part of policy makers and political figures to the goal of somehow eliminating what we call poverty. The War on Poverty, conceived by President Kennedy and launched by President Johnson, was the political outcome of researches conducted by American scholars over a generation's time on a complex of problems revolving around the fact that there continues to exist in American society a small but significant proportion of families who are socially and economically disadvantaged.

The concern with poverty apparent in the work of these scholars, and in the work of applied professionals such as social workers, has generally been manifest as a concern with those families who live below certain levels considered to be minimally adequate for subsistence. The poverty standards that have grown up in connection with the calculation of social welfare budgets and the like have been based on the assumption that there is a certain amount of income that is necessary simply to subsist. Indeed, these budgets have often been in large part determined by the cost of food for a family. (Since it is known that families at low-income levels spend about one-third of their income on food, the subsistence level has been calculated simply as three times the amount of money that is necessary to provide a minimally adequate diet for a family of given size.) Yet we now know that these "subsistence income levels" have tended to rise faster than has the cost of living, so that what would be considered a subsistence level income today is far above the subsistence level income of twenty or forty years ago even when price increases are taken into account.

All of this suggests that what is called poverty in the United States is a relative matter—relative to time and place and how well off the rest of the population is—and not a question of some absolute level of subsistence that is unchanging over time. Once this point of view is adopted the dimensions of the poverty problem are quite different. While, in terms of an absolute standard, poverty has diminished dramatically in the United States, in terms of any reasonable relative definition of poverty there has been little if any change. For example, taking today's $3,000 poverty line, over the past twenty years the proportion of families who would be considered poor has been cut in half.

But this kind of economic definition of poverty has never really been central to the things that concern Americans, intellectuals and men in the street alike, about social and economic disadvantage. In fact it has been the social problems which seem to accompany poverty that have perplexed and frustrated the nation. One could well argue that it really matters very little if some people continue to earn a great deal less than the average family income so long as they are doing slightly better each year economically, and so long as in fact most of them have the opportunity to acquire reasonable amounts of food, have a roof over their heads, and clothes to keep them warm. The concern with the problem of poverty is motivated by the many specific social problems which seem somehow related to the fact that there exists in the society a group of people removed from the accustomed affluence of average American life. Some of these problems have to do with the fact that a significant minority of men and women, boys and girls, are not able to establish themselves within the institutions that keep the society going economically—these are the problems of adolescents leaving school before they have graduated from high school (which in the United States is increasingly regarded as the minimum adequate education), of children from these families tending not to achieve adequately in school and once they have entered the labor market tending to have rather high rates of unemployment and irregular employment.

The poor are, in short, marginal to the major training and productive institutions of the society. Another kind of problem has to do not so much with the productive sectors of the society but rather with the fact that the same substantial minority of the population seems to be disproportionately involved in various kinds of deviance from the standards presumably held for all members of the society—here we become concerned with the problems of crimes of violence and of property, with problems of juvenile delinquency, illegitimacy, physical and mental disease, divorce and separation, and of mothers with children who have no husbands and are unable to support their families adequately. Most recently all of these problems have reached a new level of expression in the widespread rioting and mass violence that has affected most of the major American cities during the 1960's.

The social scientists who have been most concerned with these social problems have not been economists, but rather sociologists, anthropologists, and psychologists. They have tended to define these problems as those of social class rather than simply as problems of income, and from their perspectives the problem of poverty appears as the problem of the existence of a lower-class group in the society. The specific social problems enumerated above—crime, illegitimacy, poor school achievement, and so forth —have been seen by these behavioral scientists as arising from the generalized situation of the lower class, each such problem being only one particular expression of a more general malaise that comes from living at the bottom of the social hierarchy. After several decades of intensive research and theoretical exploration of this complex of problems it is now possible to begin to see some of the ways in which the behavioral scientists' emphasis on social and cultural processes interact with the economists' concerns with income and income distribution.

The behavioral scientists have learned that most of the problems of those who live at the lower-class level, and the problems that lower-class people in turn make for the rest of the society, can be understood if conceptualized as a result of the lower-class person's efforts to adapt to and cope with the relative deprivation of being so far removed from the average American standard. The people who are called poor and near poor live a life somewhat separated from that of the stable working- and middle-class members of the society. The latter have generally been doing very well indeed in terms of material affluence; in many ways they have found their lives more and more gratifying as the country has progressed economically, even as the sense of alienation of the relatively most deprived has increased.

The lower class is defined by two tough facts of life as it is experienced from day to day and from birth to death. These are the facts of deprivation and of exclusion: the lower class is deprived because it is excluded from the ordinary run of average American working- and middle-class life, and it is excluded because it is deprived of the resources necessary to function in the institutions of the mainstream of American life (which is after all working, and not middle, class). The most basic deprivation is of course the lack of an adequate family income, but from this deprivation flows the sense so characteristic of lower-class groups of not having the price of admission to participation in the many different kinds of rewards that ordinary society offers, some of which cost money, but also a good many others (education, for example) that do not.

But deprivation and exclusion are only the beginning of the troubles of the lower class, and sometimes in day-to-day life they do not not really loom largest as barriers to a sense of reasonable satisfaction and security about who and where one is. The economic system and the system of social segregation operate to concentrate lower-class people into particular communities. In those communities, by virtue of their own troubles and by virtue of the indifference and exploitative attitudes of the rest of the society, there grows up a system of institutionalized pathology (to use Kenneth B. Clark's phrase in *Dark Ghetto*) which characterizes ghettos and slum neighborhoods. It is this world, more than the objective facts of deprivation and exclusion per se, that impinges most directly on the lower-class child as he grows up and on the lower-class adult as he lives from day to day.

Let me illustrate this point about the direct, immediately experienced relevance of institutionalized pathology with an example. From my own research and from a careful analysis of studies during the Depression and after, I feel reasonably certain that the high proportion of lower-class marriages which break up do so primarily because the husband is not able to be a stable wage earner. That is, they break up because of "deprivation," and the exclusion from ordinary life that follows from it. However, in a study in which we asked a representative sample of women in a public housing project why their marriages broke up, we discovered that the women themselves did not give low income or their husbands' inability to find work as the most frequent reasons for their marriages breaking up. Instead, some 50 per cent of them said that their marriages broke up because their husbands were unfaithful, played around in the streets, or drank too much, and another 27 per cent indicated that their marriages broke up because the husbands *would not* work or give them enough money. In other words, from the wives' point of view, the breakup of the marriages was the fault of the husband, and not an unwilled result of living in highly depriving circumstances.

Perhaps this is just another way of saying that lower-class

people, the same as the rest of us, are not capable of being detached and impersonal about the events of their daily lives. But this is a very important fact, because when such incidents are multiplied thousands of times in the interactions of husbands and wives, parents and their children, friends and neighbors, and when this kind of individual faultfinding comes to dominate the contacts that lower-class people have with the functionaries on whom they depend in social agencies, educational institutions, housing authorities, and the like, an institutionalized system exists in which lower-class people are constantly subjected to "moral damage" in their interactions with others. By their very peers, and even more by their "superiors," they are deprived of a right which even the most uncivilized and primitive people that anthropologists have studied routinely accord their members, that is, the right to consider oneself and to be considered by others a worthwhile and valid representative of the human race.

To me this is the central damaging fact of lower-class existence. The potential for an attack on one's moral worth is ever present for lower-class people. Depending on their particular situations they learn to expect such attacks from other family members, from their peers, from their neighbors, and most predictably from the caretakers with whom they have contact.

But lower-class people have responded to this reality of their lives with all of the resourcefulness and imaginativeness that human beings can bring to bear in dealing with their difficulties. Their ways of coping with these dangers (as well as more obvious physical dangers and socioeconomic frustrations that come from their deprived and excluded existence) work in the sense that they are effective in maintaining daily lives that are for the most part tolerable, though seldom highly gratifying. At the same time these ways of coping with the dangers with which their world presents them also have a negative feedback into the system of social relations of the group in that they sometimes precipitate further problems for the individual and set problems of adaptation for others in the group and for outsiders. When social scientists speak of the tangle of pathology of ghetto and slum worlds they refer to these negative feedback effects of the adaptive mechanisms of lower-class people, although they are also aware that the same mechanisms may well be the only ones that are meaningfully available to the individual who grows up in a lower-class world.

It is in the light of this very abbreviated assessment of the social and cultural situation of the lower class that we can assess most effectively the appropriateness of the public policy response to the problem of American poverty. In general, American poverty policy has been oriented to a minimum subsistence idea of what constitutes poverty. The main goals of War on Poverty programs have been to provide services which alleviate some of the problems that result from poverty—as in the provision of special health, legal, housing, family life, and education programs—and to develop other programs that will train the poor to behave in ways that would allow them to attain jobs which pay about the minimum subsistence level of income. In other words, the War on Poverty has sought a two-pronged attack which on the one hand changes the poor person into a more conventional person in terms of style of life and work-oriented habits, and at the same time makes opportunities available to him so that he can translate his new social and technical skills into a stable job.

The results of these programs, however, have so far been discouraging. Some supporters of the War on Poverty programs argue that this lack of success is simply a result of the token financing of the program—after all, they argue, how can one expect to eradicate poverty with an annual expenditure by the Office of Economic Opportunity of less than $100 per poor person. Among social scientists, however, there is increasing skepticism that the services and opportunity programs could achieve significant impact on the poverty problem even with vastly increased budgets.

This skepticism comes from the recognition of the two cardinal facts about American poverty outlined earlier: first, that American poverty is a phenomenon of relative deprivation and not living below some absolute minimum subsistence level, and second, that the behavior of poor people which often seems to interfere with their taking advantage of services and opportunities represents an inevitable adaptation to their situation of relative deprivation, an adaptation that will not easily be given up unless there is some very basic change in the situation to which they must adapt. Therefore, these social scientists argue that the goal of any seriously intended War on Poverty must be that of reducing income inequality by insuring a minimum income floor such that no American family has a standard of living significantly below that of the average family. The first stage of such a program would be directed at those who receive less than half of the median family income; their incomes would be increased to half the median. This program would affect about 20 per cent of American families. But it seems likely that many of the social problems associated with poverty would not be more than moderately reduced and that the second stage of an income equalization program would have to take as its goal the increasing of family incomes in the lower 40 per cent of the population so that a minimum family income floor of around $5,000 would be established.

These are formidable goals indeed, and once they have been formulated the question immediately arises as to whether it is possible, either economically or politically, to achieve such goals.

The United States' socioeconomic system is in many ways a highly stable one; so stable, in fact, that it can tolerate without serious dislocation even an ongoing low-level revolt of the disadvantaged in its cities. (For example, there was hardly a ripple on Wall Street as riots swept Newark and Detroit and numerous smaller cities.) This American socioeconomic system can and does insure an increasing affluence for the nation as a whole, an affluence that is brought home to working- and middle-class families each year as a slightly higher standard of living. But this same system also operates in such a way that a large portion of the population is doomed to live removed from the rest, and it seems by now quite clear that there are no self-corrective mechanisms in the system that will, in fact, gradually reduce the size of the population that lives in "the other America." In addition to the purely economic forces which preserve a highly unequal income distribution, racial prejudice and discrimination, both historically and currently operative, continue to reinforce inequality and frustrate most government efforts to alleviate the problem through legal measures, job training, improved educational programs for the so-called culturally disadvantaged, and so forth. Obviously, it will take a major change in the way the nation conducts its affairs and allocates its resources to alter the socioeconomic hierarchy in such a way that the problems associated with poverty and racial exploitation no longer exist. The curious thing about this change is that, were it made, it would probably result in even higher levels of affluence and social stability than now exist, but the nation nevertheless pulls back from the price that would have to be paid to bring about these conditions.

The economists and sociologists who have been concerned with this problem have suggested that the following kinds of measures will be necessary to begin to reduce income inequality. First of all, the nation's economy will have to be operated in such a way that there is tight full employment. That is, the 4 per cent unemployment rate that the United States now considers acceptable must be sharply reduced to something in the 2 per cent range. In addition, because a great many of the poor are immediately capable of functioning only at the unskilled and semiskilled levels, there will have to be specific government financing designed to create work for persons who have those kinds of skills. In fact, there is a great deal of such work that needs to be done, particularly in the public sector of the economy—in hospitals, in schools, in various local municipal services. During the Great Depression of the 1930's such programs were put into effect to solve the very high unemployment of that time, but they were regarded as temporary expedients. Now the problem is to build in government-stimulated demand for work in the public sector that will provide stable and secure jobs—year in, year out—not on a temporary basis.

Along with the provision of jobs, it would of course be necessary to see to it that wages paid for these jobs were such as to provide incomes that approximate the average American standard. This means that minimum wages in both the public and private sectors would have to be pegged to at least three-quarters of the wage necessary to earn the median family income, and that unemployment compensation payments would have to be similarly high. Such high minimum wages and unemployment compensation could be economically feasible only in the context of very tight full employment.

Finally, because a significant minority of the poor are not able to work—because they are mothers of many children or because they are disabled—it is apparent that there would have to be some guaranteed minimum income provisions applicable to families in which there is no wage earner or in which the wage earner cannot earn sufficient income. Recently, many such guaranteed income programs have been suggested, the most popular being the negative income tax and the family allowance. As these programs are examined and tested for their costs and other significant features, it is very likely that a guaranteed income plan will emerge which provides a reasonable income floor without at the same time significantly reducing the incentive of able-bodied persons to work.

A price will, of course, have to be paid for such programs. The low productivity of a good many poor workers will have to be subsidized for a number of years, perhaps for a generation. However, this price can be easily exaggerated, since the introduction of a large group of even unskilled workers would have the effect of increasing the Gross National Product both directly and in terms of the demand generated for goods and services by the now more affluent former poor. In any case, from the point of view of sociologists, whose assessment of research on the lower class emphasizes the necessity of changing lower-class situations before lower-class adaptations can occur, it seems that whatever the price, only in this way can the problem of poverty be eventually resolved. If it is resolved by this kind of income strategy, then one can expect at the very least that the next generation of the poor will have acquired through their schooling and through normal job training programs sufficient skills to earn their own way in society so that the cost would be, in the main, a onetime cost.

Along these lines, then, it seems that the intellectual problems of understanding poverty and of developing social policies that would resolve the problem over the long term are gradually approaching solution. But the solution of the intellectual problem of course in no way guarantees that the poverty problem will be solved in the real social, economic, and political world. At present the nation's creative political efforts are almost totally absorbed in trying to come to terms with the United States involvement in Southeast Asia, a problem beside which the problem of American poverty tends to pale as a contemporary challenge. It is unlikely that the challenge of income inequality will be seriously addressed by the nation so long as the Southeast Asian involvement continues at its present level.

In the long run, however, the destructiveness of American poverty (and its twin, racial prejudice and exploitation of the Negro and other minority groups) will eventually place it high on the national agenda, not only on humanitarian grounds, but also because without a solution to these problems it seems likely that the American city as we know it today will itself be destroyed.

# 18 SOCIAL PLANNING: THE SEARCH FOR LEGITIMACY

## Martin Rein

All planning must in some fashion resolve the problem of legitimacy, of what authority justifies its intervention. Weber devoted a great deal of attention to the problems of legitimacy and authority, as these applied to the leaders of political systems. He argued that rulers might claim legitimacy for their rule and members of the political system might in turn accept their claims on three grounds: tradition; exceptional personal qualities, known as charisma; and legality, that is, authority rooted in constitutional rules and laws.[1] This chapter is concerned with the sources of legitimacy for planners rather than for rulers, so other types of authority need to be considered.

Since planning must validate its intervention with reference to some accepted legitimate source of authority, it can be contrasted with revolution, which ostensibly repudiates the concept of legitimacy by seeking to replace established rules, principles, and standards. By contrast, reform and city planning both share a common ideological commitment to introduce social innovation—new ideas and new programs which will reduce or eliminate social problems.[2] I use the terms "planned change" and "social reform" interchangeably, for I believe that the professional planners serve more as advocates for some specific conception of social justice than, as we earlier believed, as disinterested explicators of choice, experts in a neutral cost-benefit analysis that enabled them to effectively separate facts and values. Of course, not all planning is directed at change, for it can be used to inhibit or postpone change as well. We are concerned here only with that genre of planning which seeks to bring about social change, and in this context the planner may by definition be viewed as a reformer.

But what makes the intervention of the reformer and planner meaningful and desirable? How is the need for innovative intervention justified and support for it secured? The problem of legitimacy arises out of the need to justify and to locate a supporting source of authority to give the planners and reformers the right to intervene. It is especially acute in American society because the reformer or planner has only limited power to implement his objectives. In this context, power can be seen as "the ability to control external and internal environments and/or to counteract the consequences of imperfect control."[3] Lacking sufficient power, he needs, therefore, to win the cooperation of other centers of power if he is to achieve his aims. If planners were able to use force and were to have extensive power, planning would take on a sharply different character and would pose a series of questions which are beyond the scope of this analysis.

The problem of legitimacy arises largely from the efforts to collect and to harness fragmented power in order to bring about planned change. Some planning organizations hope to bypass this dilemma by repudiating the mandate to innovate or to promote planned change. They define their mission as providing only a forum to help others reach agreement, through the intervention of "enablers" rather than "planners."[4] By contrast, organizations that promote planned change must seek the authority to impose limits on the freedom of other organizations. They attempt to subordinate the interests or change the function and purpose of some organizations in order to promote that elusive ideal we call the public interest. Yet as soon as such organizations are powerful enough to be effective, they are also strong enough to abuse their power. Efforts must then be developed to contain their power. In a democratic society, great restraints are placed on the centralization of power, while greater freedom is given to individual units. This generalization seems even more true in the social welfare field than in the operation of business and the economy, where the disadvantages of unbridled competition have long been recognized.[5] Yet when great injustices exist, some centralized power is needed to be able to correct them. This cannot be achieved without interfering with the freedom of individual organizations. How to reconcile the clashing demands between the power needed to limit social abuse and the power needed to reduce human suffering, with the reduction in the freedom of action of others which of necessity such actions entail is a great challenge to democratic societies. The search for legitimacy is an effort to resolve this dilemma.

A review of the experience of city planners in the United States suggests that they have relied on four different sources of authority to justify and legitimate their intervention. One approach is based on the assumption that planners have command over a technical and scientific body of knowledge which enables them to challenge the irrationality in the political process of city government, which has produced decisions based on "opportunistic bargaining among vested political and economic interests of great strength."[6] The early planning movement was based on a doctrine that what planners needed was great formal powers, independent of the political process, which would enable them to act as an autonomous "fourth power" in the government of the city. The first City Planning Commission of New York City, in 1934, had a majority of members who "were committed to the premise that the commission should be an institution of experts with an authoritative voice in the decisions of city government, yet be itself aloof and protected, without the necessity of bargaining with and making concessions to the 'politicians' and 'special interests.' "[7] Experience soon suggested that political autonomy leads to isolation and independence leads to impotence. Authority which is depoliticalized (independent from the political process and based on technical and scientific rationality) offers only the authority to propose, rather than the power to achieve. Not surprisingly, the need for new sources of authority was soon recognized.

The debate in the 1940's between Carl Friedrich and Herman Finer in the public administration literature influenced city planners and other professional groups. It centered around the issue of whether every administrative act must also entail a political consequence; if this were so, then there would be no purely technological solutions. Politics cannot be separated from administration, and hence the planner secures his authority from the politician rather than from his technology. However, the more incomplete the control over the administrative process the wider the influence of the professional, thus leaving ambiguous the role of the planner in developing policy.[8] According to this interpretation, the difficulty with the concept of planning as a "fourth power" was that the claims of the planners conflicted with those of the politicians. This debate is still unresolved because trade-offs rather than simply a choice between these conflicting sources of authority are needed. A recent analysis of this debate by Beckman took the position that this conflict of identity between the planner and the politician "can best be resolved . . . if he (the planner) is willing to accept the vital but more limited role that our system assigned to the public employee."[9] Beckman urges that the planner assist and serve the policy-maker and that the planner's "influence on public policy is achieved within the bureaucracy through competence. Planners and other staff advisers have influence only as they can persuade their political superiors . . . it must be remembered that in our system of government politics subordinates the public employee, grants responsibility and power to the politician, invests open authority in the voter."[10]

In principle, then, the planner is responsible to the elected representatives of government, and government in turn to the people. The planner who repudiates a decision of his superiors can try to persuade them to accept his opinion or he can resign in indignation. But in his role as an employer, the planner secures his authority from the elected representative who in turn secures it from the citizen.

The theory about the relationship between the planner as a bureaucrat and the politician as the representative of the electorate often disintegrates in practice. The scope and complexity of public bureaucracies makes them increasingly removed from review by elected officials. They control the information by which their competence can be challenged and they outlast the politicians who depend upon them to execute their policies. Moreover, elected officials serve the interests of certain groups rather than some hypothetical over-all public interest. In theory an aggrieved citizen can protest directly to his representative against any intrusion on his rights or neglect of his needs. In practice, in a democracy, the needs and preferences of unpopular, unwanted, and powerless groups are neglected. Politicians are committed to political survival. They respond to the preferences of the constituencies that elect them, rather than the needs of the inarticulate and hence unrepresented groups.

In the 1950's the case against the planner as a bureaucrat began to emerge in the writings of Herbert Gans, John Dyckman, and Martin Meyerson, in their studies of recreation, education, and health care facilities. They came to recognize that planning which was responsive to professional discretion and to political leadership might forsake in the process the preferences and needs and desires of the consuming population. At that time, what they seemed to be calling for was a new technology which could feed new information into the planning process, namely data about the preferences of present and potential service users; planning must be responsive to the consumer. Explicit criteria for resolving conflicts which might arise when consumer preferences clashed with the policies or values of the planners or of the established bureaucracy and elected officials were not developed. Though they taught together, the work of these three planners took a long time to be drawn together and written up into a single report, partly, I believe, because they disagreed on the political implications of a consumer oriented approach.[11]

Rapkin, Winnick, and Blank, in their monograph on housing market analysis, had developed a similar position, holding that the criteria for developing public policy should rest on the choices of users, identified through the market mechanism.[12] Turning to the ultimate consumer as the source of legitimacy for planning opened important ideological questions concerning the limits and possibilities of exploiting the market as a mechanism for assessing consumer choices. The writings of Davidoff and Reiner extended the general argument. "It is not for the planner to make the final decision transforming values into policy commitments. His role is to identify distribution of values among people, and how values are weighed against each other."[13] By the

1960's new forces emerged in the political process which gave currency and acceptability to the idea of consumer advocacy.

Some planners were urging that a new source of legitimacy be found, with the planner acting as a more direct advocate of the values, preferences, and needs of consumer groups, and especially of those groups that are politically inarticulate. Planning should derive its legitimacy from the needs of the people to be serviced. The planner could then offer his skills to a user-bureaucracy as contrasted with the supplier-bureaucracy to which planners presently offer their services.

These ideas found expression in the theory of advocacy planning, which asserts that planners can derive their legtimacy from the clients to be served. Advocacy implies argument and contention on behalf of a point of view or of a specific proposal. Paul Davidoff, in his influential article on advocacy and pluralism in planning, makes this position explicit when he explains that "the advocate planner would be responsible to his client and would seek to express his client's views."[14]

All of these writers accepted the position that planners derive their legitimacy from the preferences, choices, and needs of the consumers and clients who are affected by the planning decisions. But they differed on the implications of this position for action. A point of view oriented to clients can lead to social surveys of consumer choices, or to faith in the market as the ultimate mechanism for the expressions of choice, or to the defense of consumer rights within an adversary, rather than a market framework. Each position has its difficulties. The preferences of all individuals as revealed by rankings of values from surveys cannot be aggregated into collective preferences without violating the choices of some individuals, following the famous Arrow paradox. Planning originated as an effort to supplement or supersede the market when it failed to meet individual needs or solve the problem of externalities. Faith in market freedom and choice for users did not contribute substantially to resolving these issues of public policy. And as we have begun to experiment with advocacy, intractable problems have also emerged.

One account of the experience of an advocate planner suggests some of these formidable difficulties. First, it proved very difficult to identify the client or community to be serviced. A community is heterogeneous, and the efforts to locate a single client-organization to represent it which could also speak for the unrepresented elements of the community proved exceedingly difficult. But even if such a group could be organized, it was difficult to identify the interests of a heterogeneous community-client. However, when these interests were articulated, the planner discovered that local decision units can be parochial and even punishing to the poor with special problems, such as welfare mothers, skid row bums, gypsies, and so forth. Lisa Peattie offers the grim observation that "a consequence of giving every neighborhood in a city its advocate planner might be a general closing up of the city against the poor." But even if the problems of locating the client and identifying and accepting his interests can be overcome, the advocate planner confronts a disturbing dilemma when he discovers that "the citizen client group seems . . . to serve a kind of legitimizing function which permits the planners to represent themselves as something more than merely proponents of another opinion."[15]

And the source of legitimacy rests on the professional values to which the planner is committed, as well as the technical competence he claims. According to this formulation, city planning is a value laden profession and these values offer a source of authority, a sanction to plan. While there is surely an uneasy nestling together of expertise and ideology and while a general reluctance to act on the authority of the latter is evident, nevertheless, as the impossibility of separating values and technology is accepted, action based on values is taken. One form this has taken is the creation of a competing professional association committed to implementing different values. "It appears that the profession is being split into progressive and conservative wings: the former calling for social planning to reduce racial and economic inequalities, and the latter defending traditional physical planning and the legitimacy of the middle-class values."[16]

Increasingly planners are being enjoined to act as insurgents within the bureaucracies by which they are employed, to seek to change the policies and purposes of the bureaucracy in line with the declared value assumptions. These values are procedural as well as substantive. Decision rules such as involvement of those affected by decisions illustrate the former, and goals such as racially and economically integrated communities or the reduction of inequalities illustrate the latter.

Public opinion and official policies of the bureaucracy may be hostile to these values. The planner who acts as a rebel within his bureaucracy challenges its established procedures and policies. A declaration of open warfare forces the bureaucrat to resign on principle. As an outsider he may elect to infiltrate from without in his role as consultant or researcher. Many private consultant firms and individual planners are committed to promoting their professional ideology as well as the technology. But the planner may elect to stay and wage guerrilla warfare, choosing points where the system is vulnerable internally, or he may develop coalitions with external groups to create internal change. He may lie dormant for years when levers for change are absent. Yet every bureaucrat has these insurgents who are ready to act as guerrilla-reformers to shake up the bureaucracy. In the planning field, there has been at least one effort to create a loose coalition of insurgents who are seeking to change the policies of their own and other programs. Leonard Duhl calls this coalition a "floating crap game." The analogy is misleading, though, because the players are not in competition with each other, but rather seek to support each other in their common mission. Professional associations such as Planners for Equal Opportunity (PEO) may also serve as such a reference group. Duhl, a member of the faculty of the Department of City and Regional Planning in Berkeley, may be influential in teaching some students of city planning to define their roles as guerrillas, whose authority to intervene rests on their professional values.

This source of legitimacy poses awkward issues of the boundaries between professional and personal values, ethics of means and the ethics of ends to be adopted, and the procedures of professional accountability to judge when ethics has surrendered to expediency. But despite these and other difficulties, planners who repudiate the position that values and technology are separable are experimenting with this source of legitimacy.

Perhaps, in time, a literature may emerge which probes the intellectual and political viability of this source of legitimate authority. Duhl suggests that had Weber lived he might have developed this line of argument further in later attempts to refine his classic essay on legitimacy.

Even this condensed review of the history of physical planning makes it evident that city planners have sought different sources of legitimacy: in scientific expertise, independent of the political process; as agents of elected political representatives and the bureaucracies which are accountable to them; as translators and advocates of the preferences of user-groups; and finally, as implementors of professional values. Each source of legitimacy has its characteristic difficulties, as this brief review of the experience of physical planners as experts, bureaucrats, advocates, and insurgents has suggested.

But why must planners be forced to choose among these alternative sources of legitimacy? The position of planning could be substantially strengthened if it could simultaneously call upon professional technology, values and standards, established political power, and the needs and wishes of client groups as their sources of legitimacy. If these sources of legitimacy conflict when pursued together, the planner must then choose one or another source of legitimate authority. A review of the experience of what Reston called the "new breed of antipoverty planners" committed to preventing delinquency and eliminating poverty helps to illuminate the problems which arise in adapting each source of legitimacy and in pursuing multiple sources of legitimacy which are in conflict. Federal legislation such as the Juvenile Delinquency and Youth Offences Control Act of 1961 and the Economic Opportunity Act of 1964 provided the resources which made this form of social planning possible. More recently, through legislation made possible by the Demonstration Cities and Metropolitan Act of 1966, the style of planning has been extended to cope with problems of deterioration in the urban environment. These social planners and social reformers have searched for a relevant form of legitimacy, which is strikingly parallel to the search for legitimacy among city planners. This chapter then, examines the sources of authority that antipoverty and social planning have pursued, in an effort to resolve the problem of legitimacy.

The city planner and this type of social planner share much in common. They both spend a great deal of their energies writing proposals in an effort to secure federal and state funds, they are both concerned with developing specific programs to implement ambiguous and ill-defined social objectives, they are both committed to drawing up both long-range and short-range plans, both are in principle committed to the introduction of new ideas and to generating social innovations which can lay the foundation for further experimentation, and finally, both hope to have the plans they develop implemented administratively. But in the context of this discussion, the important common ground they share is the search for legitimacy. How can their intervention be justified?

In the remainder of this study, three strategies are examined from the perspective of how they contribute to resolving the problem of legitimation of reform.[17]

Each strategy is crucial. None is sufficient by itself, for each has inherent limitations, but the efforts to pursue more than one strategy at a time often lead to conflict and contradiction. Thus in the effort to resolve one dilemma another is created.

## THE CONSENSUS OF ELITES

One way of justifying intervention is to have it endorsed and supported by the leadership of the major institutions in the community. This strategy acknowledges the power of established institutions. One version seeks to influence power by boring from within, by co-opting the institutions to serve its purposes. The endorsement by established power legitimates the efforts of reform and change.

At one time in the social services this power of change was vested in the coalition of voluntary institutions that represented the elite of the community.

Welfare services and planning became recognizably controlled by an

essentially elite leadership in each community. . . . Associated with
the socially elite were an economic elite. . . . These economic sinews
became the foundation for the support of much of social welfare.
It was only later recognized that this elite leadership was primarily
white and Protestant, representing the early stratification of American society.[18]

These economic and social elites often rejected the role of
government in welfare. They believed that democracy depended
on voluntary social institutions. They held an elitist view of
democracy, in that they assumed that they were best able to
comprehend, to represent, and to protect the interests of the
"total community." Early social welfare planning as represented
by health and welfare councils placed its claim for legitimacy
in these voluntary and often antigovernmental institutions.

Today, because of the changing role of government and the
development of new centers of power, such voluntary bodies
can no longer provide an adequate base from which the legitimacy of change can be claimed. Consequently, planners have
had to seek the participation of city government by forming a
coalition of departments such as welfare, recreation, and police
or to attempt to form a coalition among units of government
such as the city, the county, the school boards, and the states.
The influence of nongovernmental elites could not altogether
be bypassed. Legitimacy depended on bringing together a broad
coalition of groups which represent old and new sources of
power—influential leaders, established organizational interests,
and government. Unlike the early planners in the voluntary
field, governmental bodies played leading, although by no
means exclusive, roles. The support of the mayor was seen as
essential to success. And, also, unlike the earlier planners, the
creation of a local coalition of elites was nurtured by national
elites. In a penetrating analysis of the relationship between
sociology and the welfare state, Gouldner calls attention to "the
manner in which social reform in the United States has changed
in character. What is new is not the 'plight of the cities,' however increasing their deterioration, but rather that . . . the
locus of reform initiatives and resources is increasingly found

on the level of national politics and foundations, rather than in the political vitality, the economic resources, or the zealous initiatives of elites with local roots."[19]

A broad-based representative organizational structure that serves to legitimate reform conflicts with its very purpose—the search for innovation and change.[20] The greater the diversity of institutional interest that is embraced within such a planning structure, the greater can be the claim for legitimacy, since it can be established that most of the total community is then represented. But as legitimacy is strengthened, innovation is forsaken in favor of maintaining a consensus on which these divergent interests can agree. These new planning structures are continually beset with internal insurrection. Involvement of community leaders does little to resolve the problems of jurisdictional conflict; indeed, it may only aggravate the task. The voluntary planning bodies and the elite community leaders who represent them want more influence than they receive as only members (rather than convenors) of the coalition. Whereas they once enjoyed pre-eminence in the area of planning, they have now been cast aside and relegated to a secondary role by this new, more widely representative, structure. Consequently, much of the energy of the planning organization is directed away from promoting innovation and change and toward solving the more intractable problems of sheer survival—the maintenance of the coalition.[21]

The national reformers who made available the funds for these local planning organizations recognized this dilemma, but hoped that it could be solved. Essentially, their strategy rested on two related assumptions, that a marginal increase in funds can stimulate change and that the involvement of voluntary and public bureaucracies is a necessary precondition for change. They hoped the power of federal money and the process of participation would lead to change. They assumed that financially starved institutions would be willing to make changes in their operation in order to secure available and needed funds. Local reformers operating from these local planning organizations would play a central role in this process, for they could serve as interpreters of the institutional changes that would be required if funds were to be forthcoming. The professional reformer, enjoying a monopoly of knowledge and special access to these nonlocal funders, could assert a substantial amount of influence on the direction in which the local coalition of elites would develop in order to obtain the wanted funds. The fundamental premise in these negotiations was that because funds were so desperately needed by local institutions, they would be willing to participate in self-reform in order to secure them. That the institutions might both obtain the funds from established and new sources and resist change was a contingency to which local and national planners seemed to have given little attention.

This second assumption rested on the faith that the process of involvement and participation must lead to self-education and the acceptance of the need to change. The implicit theory of bureaucracy on which this belief rests is that the sources of organizational rigidity are largely ignorance and faulty communication. If a context were provided in which institutions could more freely communicate with one another, the validity of the need for change would more readily be recognized and accepted. Such a theory is, however, incomplete, for it denies the existence of the more fundamental conflict of values and interests among institutions which more open communication might serve to exacerbate rather than to alleviate. Infiltrating established power rests on inadequate assumptions about how institutions perform. The more the coalition is in danger of falling apart as a result of these challenges, the more compromises are made to assure survival and the more innovation is sacrificed to achieve consensus.

Much as a representative structure reduces innovation, participation by institutions in their own reform leads also to incoherence and contradiction. The funds made available by the planners are simply not large enough to finance major reforms in these institutions. But even if sufficient funds were available, it is hard to see how the planners could conceivably hope to initiate major structural changes unless the changes were in conformity with the institutions' prevailing interpretation of their function and reflected a direction that the institutions were already prepared to take. Under these conditions it is the planners who have been co-opted by the institutions. Involvement, although it facilitates legitimation, impedes innovation.

An alternative to the theory that institutions change with self-education is derived from the assumption that institutions must be challenged, for they will not change of their own accord. The public health movement, which was promoted largely by lay groups, arose (with the support of some professionals) above the vigorous opposition of physicians; similarly, the Charity Organization Society (the mainstay and bulwark behind social casework before the Depression) was opposed to pensions for widows and the aged in the early part of the century, and the charity school opposed public education. According to this view reform cannot depend solely upon the willing cooperation of the institutions to be reformed.

While it is useless to ignore the realities of established institutional power, a program of planned change runs the dire risk of losing its sense of purpose if it relies solely on established leadership. Increasingly, planners find that the more they work with established institutions, the more compromises they have to make and the more difficult it becomes to ensure that funds are spent for innovation rather than for the expansion of the status quo. Because of this frustration, there is a tendency to bypass the major service institutions. For example, we find in education the development of preschool programs, after school programs, summer school programs, tutorial programs—all the time going around the heart of the school's mission—the everyday teaching—and creating instead a whole series of special remedial programs. Remediation becomes a kind of index of the failure to achieve more basic structural change. It represents a response to the more frustrating task of directly influencing the essential functioning of the institution itself. Institutional resistance leads to program proliferation.

Many vocational training programs have ignored the state vocational training system and have worked through other agencies or set up their own agencies to provide special programs for work training, as in the case of the Job Corps. Here duplication and remediation do not represent a retreat but rather a strategy of confrontation. But to challenge institutions so frontally, a different source of legitimacy is necessary.

## THE POWER OF KNOWLEDGE

Another way of legitimizing planned change is to offer reforms as rational, coherent, intellectual solutions to the problems that are being dealt with. This tends to be the academic approach and the approach of the professional consultant. Knowledge in the rationalistic-scientific tradition in general, and especially that which is derived from empirical research, can provide a basis for legitimacy because, presumably, it can yield valid solutions. These, in turn, depend on a value-free social science capable of objectively probing the etiology of social problems and presenting programs for action based upon fact rather than upon institutional or other value biases. The analysis of social problems and the remedies proposed for reducing or eliminating them are viewed as technical rather than ideological issues. The President's Committee on Juvenile Delinquency was especially committed to the importance of rational analysis as the basis for planning and program development, but other programs are also guided by these ideals. As a condition for receiving funds, national reformers required that communities attempt to conceptualize the problems of delinquency, poverty, or physical and social decay in the light of relevant data and social theory.

Reform stakes out a claim for legitimacy when it is based, not upon political consensus or ideological bias, but upon the hard, dispassionate facts provided by a rigorous social science analysis. Not only are the plans to be based upon a thoroughgoing, objective appraisal of the social problem, but the efforts at solution themselves are to be rigorously evaluated. With ruthless disregard for bureaucratic interests, those programs judged to be successful would be continued, whereas those falling short of the objective standards would be rejected and discontinued. Change is to be based not upon fads and vested interests but upon the evidence provided by evaluative research into program outputs. Science, rather than elitism, justifies the intervention sought by the reformer.

This strategy of reform has its own inherent contradictions. Not perhaps in the long run, when researchers can always justify their activities as ultimately contributing to truth and knowledge; but in the short run, it does indeed involve conflict, for gathering information is not without its costs. Consider the difficulty that many planning organizations have encountered in their efforts to study the conditions and problems of the Negro urban ghetto. Research, as one angry account has put it, can be a "transparent dodge for the postponement of action, that those involved in the charade of research into the problems of disadvantaged youth are willing or inadvertent accessories to those who seem to perpetuate the clear and present injustices."[22] A disillusioned Negro community wants authentic action, not rhetoric, promises, or studies.

Although these resentments may not be altogether rational, they are surely understandable, particularly when we recognize that the preliminary research and analysis of so many of the community action programs become extremely esoteric, and in many cases never really issue into pragmatic proposals at all. Indeed, it is often difficult to find coherence among the social theory, the facts presented, and the programs that are developed to reduce the problem. This widespread disjunction between the programs of reform and the research and theoretical insights represents an important limitation on the contribution of research and theory to the reduction of social problems. Much research has not yielded new knowledge about the poor, nor has it yielded especially new insights into our understanding of delinquency, nor has it led to the kinds of new programs that need to be developed to tackle these problems. The contribution of value-free information to the development of social policy has been greatly oversold.

The rigorous testing of experimental action programs has also encountered fundamental obstacles. It is difficult to include in most experimental social action programs the rigid controls that are necessary to provide the kind of clear-cut findings upon which it is possible to accept or reject particular techniques of intervention.

When this rationale of reform calls for comprehensive action involving many interrelated programs, a difficult research task becomes even more discouragingly complex. The demands for action are such that planners need to be somewhat opportunistic, flexibly adapting to the shifting political coalitions which alter substantially the content of the program on which their comprehensive program rests. But when the input variables are subject to significant change, the research task becomes even more tangled. As the limitations of the research design become even more apparent, it becomes hard to know what caused the measured outcomes. As a result, the interpretation to be drawn from the findings is open to serious question. Staunch supporters of particular programs are more likely to reject the research methodology and repudiate the criteria of evaluation, rather than to accept the implications of the negative findings that most evaluative research tends to yield.[23]

A strategy that relies upon the power of knowledge has other inherent limitations as well, for it can also conflict with the other strategies of change. Research requires a degree of autonomy if it is to follow a problem rather than yield to political expediency and feasibility. But the ruthless pursuit of a problem without regard to the question of implementation may lead to a solution that, while it is rational, is not politically relevant. This, of course, is the fundamental dilemma of all rational planning, the attempt to reconcile the conflicting requirements of rationality and feasibility. Planning that disregards the question of implementation languishes as an academic irrelevancy; it may be right but not relevant, correct but not useful. While planning and research require close integration, they make competing claims for resources. Enterprising researchers have been able to secure a very substantial portion of total budgets available to planning agencies, while unwary planners are left with reduced resources to carry out their tasks. Irrelevance can arise not only in the competition for resources, but in the conflicting value biases of the researcher and the reformer—the different emphases they give to knowledge and action. Research can become preoccupied with a spurious rigor, leading to a kind of dust-bowl empiricism which provides data without theory. Bewildered planners are left with a maze of tables and data which yield no immediately coherent themes and which provide little information from which implications for action can be drawn.

Even more than being a costly irrelevance, research can subvert the goals of the reformers. For example, the methodological bias of researchers can lead the reformers away from their original commitment to institutional change and toward a redefinition of the problem in terms of individual rehabilitation when this redefinition of the problem occurs. Research can inadvertantly convert radical planners into administrative planners. Reformers often hope that researchers will guide the development of planning policy by conducting studies that will help confirm or reject the basic underlying rationale of the organizations. Yet, to develop new programs requires social inventiveness, to which formal research appears to contribute little. The experience of the delinquency prevention program offers an interesting example. The rational reformers sought programs that would test the assumption that social institutions throw up barriers and block access to achievement, a process that contributes to increased disengagement and deviancy on the part of the rejected populations. Yet, strangely enough, social scientists sometimes act to reinforce those pressures on planning organizations which prevent them from focusing on community institutions as their prime targets of change. Researchers usually favor the more rigorous, traditional, and tested approaches of their disciplines, such as surveys of individual attitudes, self-perceptions, role models, and so forth. The indices they have developed to measure the impact of demonstration programs have been occasionally behavioral, but more often attitudinal, including both reduced rates of delinquent behavior and changing self-images. They have avoided indices that would measure institutional change, neglecting institutional analysis in favor of a more individualistic approach.

Research may conflict not only with the purposes of reform, but also with the search for elite consensus. Organizational studies that lead to the documentation of bureaucratic rigidities and social injustice may conflict with the efforts to promote cooperation and to secure consensus among the institutions that are being researched. If research relentlessly pursues data on the operation of the bureaucracy, it will uncover findings that could become a source of embarrassment to these cooperating institutions. Indeed, where such information is available, it is extremely awkward to know exactly what to do with it. If the information becomes public knowledge, it would only antagonize the institutions whose cooperation is so desperately sought. Yet maintaining secrets is always hazardous. This may account, in part, for the fact that planners have rarely insisted that researchers study institutional performance.

Finally, research may not be able to answer the problems posed by the reformer. Consider briefly one such question which local, gradual, and comprehensive programs must confront. The hub of a comprehensive social welfare program can be developed around many institutions—the outreach school, the welfare department, the health department, the mental hygiene clinic, and the employment centers. Some of the traditional voluntary services—the settlement house and the welfare councils—are no longer regarded as adequate focal institutions for the coordination of local service programs. But do we have any factual data that can guide us in the selection of one or another of these institutions as the appropriate focus for a comprehensive community program? Should programs center around health, housing, employment, education, or reducing dependency? If all of these are legitimate, must we then abandon the search for a truly comprehensive program, and settle for the present muddle of coordination, staturation, concerted services, and neighborhood centers, where the task of intersectoral planning is a rational ideal that is intellectually vacuous and bankrupt?

Just as a broadly representative planning structure can subvert innovation in order to preserve the frail coalition of conflicting interests, so too can researchers subvert the reformers' mission if they become preoccupied with methodology and use their studies to promote their professional identity rather than the interests of the action program. Concern for rigor and professional identity may lead to the neglect of relevant action problems. Concern for knowledge without explicit, carefully developed ideology contributes to narrow technicism.

## THE POWER OF THE PEOPLE

Reformers can also claim legitimacy if their programs are endorsed, supported, and created by the recipients of the services themselves. Such an approach has the advantage of avoiding the arrogant assumption that the technical expert or the elitist best knows the needs of the poor. It avoids the onerous charge of welfare colonialism or paternalism, wherein one group in society provides services on behalf of another. Recipients of the service are defined as politically articulate consumers, as citizens rather than as clients in need of therapy and care. Democracy is, after all, not only the search for elite consensus but also the mobilization of interest groups, each striving to pursue its own aims in the context of a pluralistic society. The American democratic system, according to this view, depends on rectifying "the basic imbalance between elites and non-elites by modifying the power differential between them."[24] It attempts to carry out this strategy by providing disadvantaged groups with more powerful instruments for articulating their demands and preferences. It helps them to organize protests in which their moral claim to justice and equal treatment can find expression. In addition to collective action it places before the poor the machinery of law through which they can act as plaintiffs against the institutions that have bypassed their rights.

Strategies of planned change which derive their legitimacy from the direct participation of local citizens and service users have had a stormy history since they were launched by the President's Committee on Juvenile Delinquency and Youth Crime in 1962. These developments must be seen in the context of the civil rights revolution and the emergence of militant demands for black power. In response to pressures, representative community-wide structures were broadened to include those individuals and groups which were the targets of change. The principle of "maximum feasible participation" under the Economic Opportunity Act was administratively interpreted in many ways, including the direct participation of the poor on the policy-making boards of Community Action Agencies.[25] In some cities, such as San Francisco (and later Oakland under the Model Cities Program), participation was interpreted as control, as the poor dominated the board with the mayor retreating to a subordinate role.[26] As organizational resistance to social change was encountered, participation turned to social protest and social action which took the form of rent strikes, boycotts, picketing, and other strategies of confrontation to promote change.[27] More recently, citizen participation has come to mean community control of social services, such as multi-service centers, health programs, and a decentralized public school system.[28] The Model Cities Program encouraged experimentation with advocacy planning, where local groups were given resources to hire their own planners to enable them to develop plans which incorporate social and physical resources for the reduction of urban blight. The Community Self-Determination Act of 1968, now before Congress, may yet usher in a new phase of user control, for it is designed to create community-controlled business enterprises that would permit the people of the community "to utilize a share of the profits of (these) enterprises to provide needed social services."[29]

In the following comments I wish to explore some limitations of this strategy as I have come to understand them in my role as consultant to community action and model cities programs at the national and local level.[30]

This strategy, like the others, has inherent limitations, and it also conflicts with the other strategies. The acceptance of this argument leads to an anomalous position, for it inadvertently supports a different interpretation of democracy for the poor than is held for other segments of society. In the middle-income style of democratic involvement, citizens work through their representatives; whereas in low-income communities, democracy tends to be interpreted as a form of direct participation at the grass-roots level. Community competence through self-help becomes defined as a therapeutic process for promoting social integration. Competent communities produce competent men, as each man is his own politician. Organizations are expected to develop spontaneously out of the mutual interests of residents working side by side on common problems. The rewards of participation are defined as civic pride, personal growth, and the reduction of community deviancy. The groups are not forged out of the more pragmatic interests in personal favors and economic advantages which more typically characterize the motives of those who join local political parties. The task might better be defined, not so much as increasing the competence of low-income communities to manage their own affairs, but rather as creating more representative structures which will be more responsive to the special needs and interests of low-income groups. Paid politicians rather than paid community enablers may be necessary if representative rather than direct democracy is to be achieved. Direct democracy may thus be seen as a stage in the process of developing new political coalitions and new political leadership rather than as an ideal in the good community.

Richard Cloward and Francis Piven have argued that a fundamental conflict between elite and low-income collective protests arises because they are based upon "quite divergent beliefs about the nature of social, economic, and political institutions and what it takes to change them." The elitist approach assumes that institutions change by persuasion and education and that issues are largely technical and capable of analysis in terms somewhat analogous to a cost-benefit evaluation. Low-income collective protests, by contrast, view institutions as responsive to naked power and pressure; issues are defined in personalized terms; opponents are seen as culprits; and exploitation of the poor is rejected, whatever the benefits.[31]

Efforts to organize low-income communities encounter difficulty in sustaining a high level of interest and participation, especially when programs have only marginal meaning for the residents and offer little opportunity for changes in jobs, housing, or other amenities. There is also the danger that issues are selected more for their capacity to rally interest than for their intrinsic merit. Protests can become ends in themselves instead of platforms for bargaining and negotiating. But without an issue for protest, organizations are likely to succumb to the meaningless ritual of organization for its own sake. Saul Alinsky's work is a prototype of one approach that attempts to sustain commitment by polarizing a community around an issue and then ruthlessly attacking the villain who is alleged to have created the problem in the community.

While it has been exceedingly difficult to organize the poor on the basis of their poverty or social class in the occupational hierarchy, there is at least precedent for politicalization along ethnic and religious lines. Citizen participation reflected this aspect of American political life and extended it by inadvertently becoming a program for organizing the Negro community. The more militant the black urban action programs became, the more the militants discovered how difficult it was to bring about change. The heightened sense of relative deprivation converted the process from reform (defined either as therapy and self-help to achieve a competent community or as opening opportunities through organizational change to promote mobility) to revolution, which found expression in riots and the repudiation of integration as a realizable social ideal. How established power will respond to violence and assault on its citadels is unclear. There is evidence of both a backlash and an increased liberalizing as the desire for social stability and social justice is joined and divided.

Lipsky has pointed out that "protest groups are uniquely capable of raising the saliency of issues, but are unequipped —by virtue of their lack of organizational resources—to participate in the formulation or adoption of solutions to problems they dramatize."[32]

When protest groups are sponsored by social welfare organizations they rapidly lose their authenticity as grass-roots movements. They drift into laborsaving self-help projects and cleanup and fixup programs. It is rather startling to note how many bureaucracies attempt to organize low-income residents as a device for promoting bureaucratic goals. Sanitation departments create block groups, the settlements organize neighborhood councils, the schools promote P.T.A.'s, and urban renewal in a similar fashion attempts to mobilize a community as a device for co-opting and reducing opposition to renewal plans. The professional comes to plan the agenda, and when the professional leaves, the organization collapses.

The process of involving the poor as a form of therapy and self-help on the one hand, and legitimization of the activities of the planners on the other hand, does not take adequate account of the potential role that citizen participation may have in politicalizing the poor. It can serve as well to create a new center of power by revitalizing the urban political machinery in low-income areas, replacing the atrophying structures that once helped adjust generations of immigrants to American society. It is paradoxical that the targets of reform in one generation should become the ideals of the next generation: ethnic politics and the political machine were once seen as the major impediments to good local government. However, many of the groups that do participate in these "establishment"-sponsored programs are suspiciously regarded as having "sold out" their allegiance to the community from which they came. But this harsh assessment of betrayal fails to recognize that involvement, when seen from the point of view of its consumers, is a way of "buying in" on a system they aspire to be part of.

### DILEMMAS IN THE SEARCH OF A SINGLE SOURCE OF LEGITIMACY

We have described, then, the three strategies that reformers and planners rely on to legitimate their action. Each appeals to a different aspect of the democratic process: the need for consensus among elite institutional interests, the reverence for science and fact, and the validation of pluralism, diversity, and conflict on which democracy depends for its vitality. The dilemma seems to be that the reform that works with the establishment, searching for a consensus, tends to lose its soul and its purpose. It abandons its real feeling and commitment for the poor as it sacrifices innovation and reform for survival and growth. Yet any program that is based solely on a fight for the rights of the poor and that fails to work with established institutions as well not only is likely to create conflict, but also may fail to generate any constructive accommodation which can lead to real reform. Organizing the poor on a neighborhood basis cannot achieve very much by way of fundamental change. Vision is limited to issues around which local initiative can be mobilized; most typically there is failure to give attention to broad social and economic policy. Research can interfere with both functions, for it can be used, in Gouldner's graphic term, as a "hamletic strategy" of delay and procrastination, responsive to political realities, while avoiding action that will provide authentic services for the poor. Research can compete with reform for resources, and it may pursue competing aims. The documentation of social injustice, which seeks action by confrontation, may embarrass the bureaucracies and make cooperation with the reformers more difficult. But without research, without some kind of objective analysis of the consequences of action, social policy moves from fashion to fashion without ever learning anything. It is, after all, useless to continue to create innovations and to spread new ideas if one never checks to see whether the new ideas and innovations are mere fads or whether they do indeed produce any kind of demonstrable change.

How then can these dilemmas be resolved? The answer, I believe, is that they cannot, for the contradictions are inherent in the nature of American social life.[33] It is futile to search for paradigms and prescriptions that will clear the whole problem out of the way and will ultimately demonstrate that the strategies are indeed consistent and mutually reinforcing, rather than fundamentally in conflict. The search for a welfare monism that rejects pluralism and conflict only fosters utopian illusions. When all three strategies are pursued simultaneously in the same organization, internal conflict develops over time. Eugene Litwak has suggested that we typically resolve such conflicts by having the conflicting functions carried out by separate organizations.

A society might stress both freedom and physical safety. These two values may conflict . . . yet the society seeks to maximize each. One way of assuring that each will be retained, despite the conflict, is to put them under separate organizational structures; that is, have the police force guard physical safety and the newspapers guard freedom of the press.[34]

Fragmentation of function does not, however, resolve the dilemma; it serves only to exacerbate the problem of inter-organizational relationship as the lack of coordination becomes a perpetual crisis.

The government's delinquency prevention program stressed rational planning and the power of knowledge, the antipoverty program sought to implement the ideals of "maximum feasible involvement of the poor," while the Model Cities Program seems to direct its energies, at least initially, to established power. This does not imply that each of the programs neglects the other strategies, but they do indeed seem to emphasize one at the cost of the others. Typically, then, we pursue all the strategies in the same and in different organizations; but also, at different points in history we stress one or another of them. We move to research when we become particularly conscious of a lack of knowledge, a lack of clear ideas. We move toward advocacy and direct action to help the poor when we are aware of the extent to which programs seem to become en-snarled in or captured by established power. We move toward established power when we feel that organized programs, what-ever their merit, fail to keep pace with the changing conditions of society and have developed barriers that prevent services from reaching those for whom they were designed.

## CONCLUSIONS

Physical and social planners have proceeded under the as-sumption that the consensus of values which binds together a society offers the most compelling frame of reference for a "community regarding" planning process. However, when the divisions which separate society become evident and the chasms which divide its groups become deep, planning at all levels comes to reflect the condition of the society for which it plans. And although the need for disinterested planning becomes more urgent when the disharmonies in the society become more evident, it also becomes equally more difficult to perform the task of harmonizer and integrator. Rational planning is a myth when the value consensus on which it must depend is illusory and the technology for eliminating arbitrary decisions is not available. But as the conditions of society become more complex and as each decision is a response to short-range expediencies and accommodations to conflicting vested interests, the need to protect the long-term interests of society becomes more insistent and the demands for rational solutions more urgent. But the absence of consensus and the limits of social science multiply the sources of authority by which planning can legitimate itself. The crucial dilemma of planning cannot be altogether avoided: the social problems which beset our society require disinterested, rational, and politically independent solutions. However, we have no technology which lends itself to objective assessment, nor have we or can we ever devise a way of detaching planning from political pressure, without at the same time converting the detachment into irrelevance. Nor is advocacy planning a solution, for one planner as advocate implies yet another (not necessarily a planner) as judge. A judge is not simply a mediator among conflicting interests; what makes his decisions just is that they conform to some normative standard, some moral value judgment.[35] But society has not thrown up mechanisms of adjudication nor has it created a body of law and tradition which provide us with the norms and standards to judge social policies when they conflict. This situation arises because the effect of social policies tends to be distributional, in that they leave some groups better off and others worse off. Social planning then, impels us to go beyond Pareto optimality as a criterion for public decision-making. Even the choices of means are never neutral insofar as ends are concerned. Because the tools for intervention embody value goals, no simple calculus for distinguishing means and ends is at hand. As a result, social science alone cannot help us choose among conflicting goals, nor can it offer criteria when public policy requires interpersonal comparison of utilities, nor can it offer criteria other than efficiency and effectiveness in those empirical studies which try to bring together means and ends.

Since no resolution of these fundamental dilemmas is at hand, each source of authority which legitimates planning offers as well an alternative interpretation of its role. Thus planning can be interpreted as disinterested planning, which seeks to exploit whatever available consensus is at hand and to plan in terms of these areas of common agreement. Planning is then interpreted as a rational scientific process in which, when goals are known, the relative efficiency of various means can be assessed. Alternatively, as Reiner has suggested, the planner can be a rational goal technician, explicating the muffled goals among choices already made on other grounds.[36] Or, when agreement is lacking, planning offers a forum from which the planner tries to forge harmony among conflicting interests. Or the planner may be seen as a bureaucrat accepting his goals from the interpretation of his mission by elected officials. But when established political patterns are rigid, and reject or exclude groups, the planner may then act as an advocate, organizing community groups to enter into the political process. Or he may serve as a guerrilla attempting to initiate change in bureaucracy by enhancing their competence and responsiveness: he himself having no explicit agenda of reform other than the wish to be relevant to current social problems.

From this review of the problems of legitimacy in social planning, two general conclusions can be reached. (1) The source of legitimacy in planning is neither self-evident nor narrowly restrictive. Indeed, there are multiple sources of legitimacy. However, they cannot all be pursued under the auspices of one planning organization, and hence choice is required. Planning need not rely solely on established elites for its legitimacy, as was the case in the past. Other sources of authority are available. (2) Each source of authority has its characteristic weaknesses and strengths which present to the planner a set of intractable problems which are moral in character and from which there can be no retreat into technology. These issues become explicit when the implications and consequences of choosing among the various sources of authority to legitimate planning are understood.

Schools of city and social planning should not organize their curriculum around a single source of legitimacy—physical planning around technology and social planning around advocacy. We must better prepare students to understand that there is a far wider range of choices which can legitimate their work and role. But equally so scholars should probe more deeply the various dilemmas of planning when each source of legitimacy is accepted. As for educators, they should help their students understand the moral problems posed for planners in a pluralistic society. This is a matter of process as well as substance.

## Notes

1. Max Weber, *The Theory of Social and Economical Organization,* trans. A. M. Henderson and Talcott Parsons (New York: Oxford University Press, 1947), p. 328.

2. In its origins American city planning drew inspiration from the civic reform movements of the time.

3. David A. Armstrong, "Some Notes on the Concept of Planning," (London: Tavistock Institute of Human Relations, July 1964), mimeographed p. 8.

4. For an analysis of the dilemma that such organizations confront when they seek to promote change, or when they fail to embrace in their forum all relevant community interests, see Martin Rein and Robert Morris, "Goals, Structures, and Strategies of Planned Change," *Social Work Practice, 1962* (New York: Columbia University Press, 1962).

5. Gabriel Kolko, *The Triumph of Conservatism: A Reinterpretation of American History, 1900–1916* (New York: The Free Press, 1963). The author argues that at the turn of the century the stability and control of big business were threatened by competition from an increasing number of small firms. After an unsuccessful effort to limit competition by voluntary means, big business turned to federal regulation. Thus business captured politics in order to introduce regulations needed to promote economic growth.

6. Wallace S. Sayre and Herbert Kaufman, *Governing New York City: Politics in the Metropolis* (New York: Russell Sage Foundation, 1960), p. 372.

7. *Ibid.,* pp. 372–373.

8. Carl Friedrich, *Constitutional Government and Democracy* (Boston: Ginn & Company, 1946); and Herman Finer, *Theory and Practice of Modern*

*Government,* rev. ed. (New York: Henry Holt & Company, 1949), pp. 871–885. For useful summary of the debate see Glendon Schubert, *The Public Interest* (New York: The Free Press, 1960), pp. 120–121.

9. Norman Beckman, "The Planner as a Bureaucrat," *Journal of the American Institute of Planners,* XXX, No. 4 (November 1964), p. 324. See also Alan Altshuler, *The City Planning Process: A Political Analysis* (Ithaca, N.Y.: Cornell University Press, 1965).

10. *Ibid.,* p. 327. Altshuler argues that since "political officials seldom give planners any clear instructions to guide the value-choice aspects of their work," much discretion remains with the experts (p. 326). He thus appears to reject Beckman's formulation of the planner's role as bureaucrat.

11. A book is now in preparation, tentatively titled, "Community Facilities, Social Values, and Goals in Planning for Education, Health, and Recreation."

12. Chester Rapkin, Louis Winnick, and David Blank, *Housing Market Analyses: A Study of Theory and Method* (Report from the Institute for Urban Land Use and Housing Studies for the Housing & Home Finance Agency, 1952).

13. Paul Davidoff and Thomas A. Reiner, "A Choice Theory of Planning," *Journal of American Institute of Planners,* XXVIII, No. 2 (May 1962), 108.

14. Paul Davidoff, "Advocacy and Pluralism in Planning," *Journal of the American Institute of Planners,* XXXI (November 1965), pp. 331–338.

15. Lisa R. Peattie, "Reflections on Advocacy Planning," *Journal of the American Institute of Planners,* XXXIV (March 1968), 84, 86.

16. Herbert Gans, "Social Planning: Regional and Urban Planning," *International Encyclopedia of the Social Sciences* (New York: The Macmillan Co. & The Free Press, 1968), p. 131.

17. The claim to a source of legitimacy must be only illusory. Hence we need to pay attention to inauthenic claims to legitimacy. Under special circumstances myths can be very important in convincing others that the claim should be heeded. This chapter does not systematically explore this important issue.

18. Robert Morris and Martin Rein, "Emerging Patterns in Community Planning," *Social Work Practice, 1963* (New York: Columbia University Press, 1963), p. 156.

19. Alvin W. Gouldner, "The Sociologist as Partisan: Sociology and the Welfare State," *The American Sociologist,* XIII (May 1968), p. 109.

20. The use of the terms "innovation" and "change" may require some explanation. I find useful the distinction developed by Lake in his review of theories and research about social change. He states, "there is no clear distinction in the literature reviewed between a *strategy* of innovation and a *theory* of change. However, that such a distinction implicitly exists seems evident from the different substance of each study." A strategy of innovation involves specific tactics for inducing change, while a theory of change "is not accompanied by a program for inducing change." Dale G. Lake, "Concepts of Change and Innovation in 1966," *The Journal of Applied Behavioral Science,* IV (1968), pp. 4–5. Kahn and his colleagues treat innovation as the procedures, roles, and activities

which enable an organization to depart from fixed rules in the face of changing circumstances. For their discussion of innovative roles within an organization, see Robert Kahn, et al., *Organizational Stress: Studies in Role Conflict and Ambiguity* (New York: John Wiley & Sons, 1964), Chap. 7. This chapter is concerned with the efforts of some organizations to induce innovation in other organizations. This process I define as planned change or social reform.

21. For further evidence about the conflict between innovation and broad-based organizations in voluntary social welfare organizations and in health and welfare councils, see Martin Rein, "Organization for Change," *Social Work*, IX, No. 2 (April 1964), 32–41. And see Martin Rein and Robert Morris, "Goals, Structures, and Strategies for Community Change," in *Social Work Practice, 1962* (New York: Columbia University Press, 1962), pp. 32–41. For a review of the literature of international organizations which reaches a similar conclusion, see Richard E. Walton, "Two Strategies of Social Change and Their Dilemmas," *Journal of Applied Behavioral Science*, I (April-May-June, 1965), 167–179.

22. *Youth in the Ghetto: A Study of the Consequences of Powerlessness and a Blueprint for Change* (New York: Harlem Youth Opportunities Unlimited, 1964), pp. 2–3.

23. See, for example, Melvin Herman, "Problems and Evaluation," *The American Child*, Vol. 47, (March 1965), and Comments on staff paper prepared by Dr. Sar Levitan, "Work Experience and Training," in *Hearing before the Subcommittee on Employment, Manpower, and Poverty, Committee on Labor and Public Welfare, United States Senate*, 90th Congress, 1st Session on S. 1545 Part 10, pp. 3072–3081.

24. Peter Bachrach, "Elite Consensus and Democracy," *The Journal of Politics*, XXVI (1962), 451.

25. For a discussion of the critique of participation as policymaking, see John C. Donovan, *The Politics of Poverty* (New York: Pegasus, 1967), pp. 41–48.

26. James Cunningham, "The Struggle of the American Urbident for Freedom and Power," a report prepared for the Ford Foundation, August 1967. See pp. 57–69 for an account of community action in San Francisco.

27. For a thoughtful appraisal of the limits and strengths of protest see Michael Lipsky, "Protest as a Political Resource," Institute for Research on Poverty, University of Wisconsin, April, 1967 (mimeo).

28. Hans Spiegel, *Citizen Participation in Urban Development* (Washington, D.C.: NTL Institute for Applied Behavioral Science, 1968), pp. 271–291.

29. *Congressional Record*, Senate, July 24, 1968.

30. For a more extensive and balanced review, see Martin Rein and S. M. Miller, "Citizen Participation and Poverty," *University of Connecticut Law Review* (January 1969).

31. Richard Cloward and Francis Piven, "Low-Income People and Political Process," a paper presented at the Training Institute Program on Urban Community Development Projects (New York: Mobilization for Youth, May 1965).

32. Michael Lipsky, "Protest as a Political Resource."

33. For a further discussion of each of these strategies and how they conflict, see Peter Marris and Martin Rein, *Dilemmas of Social Reform: Poverty and Community Action in the United States* (New York: Atherton Press, 1967).

34. Eugene Litwak and Lydia F. Hylton, "Inter-organizational Analysis: A Hypothesis on Coordinating Agencies," *Administrative Science Quarterly* (March 1962).

35. Gouldner, "The Sociologist as Partisan: Sociology in the Welfare State," p. 113.

36. Thomas A. Reiner, "The Planner as Value Technician: Two Classes of Utopian Constructs and their Impacts on Planning," in H. Wentworth Eldridge (ed.), *Taming Megalopolis* (New York: Anchor Books, Doubleday & Co., Inc., 1967), 232–247.

# 19 INTERNATIONAL URBAN RESEARCH

**Francine F. Rabinovitz**

The dominant note usually struck in discussions of international urban research is that urban development must today be thought of in global terms because cities have spread around the entire world. Within the last half century, the percentage of the world's population in cities of 100,000 or more persons has more than doubled. By 1990 over half of the world's population will live in cities of this size. Urbanization has come not only to North America and Europe but also to the developing areas. Of the world's 36 largest urban agglomerations, 17 are in developing countries. Indeed, the annual average growth rates of the urban populations of Africa, Asia, and Latin America are now from 3 to 5 per cent higher than those of the developed world.

Citing such figures about world-wide urban development seems to imply two things. First, it suggests that what matters about this process is the general growth of the urban population. Second, it suggests that urban development today, resulting from and causing processes occurring on a world-wide basis, is an essentially uniform phenomenon. But a close look at recent comparative analyses of cities indicates that a revision of the customary picture of international urbanization is required. It is worthwhile therefore to describe briefly some generally accepted hypotheses regarding the pattern and impact of urbanization. Then we will look at the alternative hypotheses that are now beginning to appear, in order to get a clearer picture of the current findings of international urban research.

Traditional interpretations of international urbanization see an association between the process or state of population concentration into aggregates in a geographical area, which we call "urbanization," and changes in individuals and nations which promote modernization in society, economy, and polity. Actually this is a two-step process. Urban development throughout the world is not only a quantitative but also a qualitative change. Shifts in settlement patterns are important because the increase in the size and number of cities affects the behavior of nations and the way of life of individuals. "Urbanization" produces a particular way of life, which we call "urbanism." It is this way of life which is said to be functional for national development.

On the individual level, the urbanite, particularly in developing areas, is more exposed to mass communications media, educational opportunities, and experiences in interacting with others than is his rural counterpart. In the city, lineage and tribal groups relinquish much of their responsibility and free the individual to make new associations. These changes do not create a vacuum, because urban centers promote the emergence of new patterns of voluntary association. Such experiences give the urban resident the equipment to make demands on the political system in a productive way. In the Middle East and Latin America, urbanization is said to have created high social mobilization. The higher the social mobilization, the greater the political participation, particularly that manifested in voting.

On the city level, urban concentrations appear crucial for carrying on many of the activities which are indispensable to nationhood in modern times. With the growth of cities, transportation and communication facilities become more efficient, allowing the exchange of services to become more specialized. As the city promotes functional differentiation it also enhances coordination among diverse efforts, for most national leadership and administrative groups are based in cities. In Latin America, even the twentieth-century peasant movements were organized by urban-based political groups or parties who sent revolutionaries out to the countryside. The big city is thus a command post for the nation, as well as a container for specifically urban activities.

On the national level, the growth of cities is associated with expansion of the gross national product, higher literacy rates, the extension of communications and medical facilities, and national political stability in Europe, Australia, Canada, and Japan. Conversely, few of the poorer, more unstable nations of the world have high proportions of population living in cities, although each does have one or two very large cities. These traditional hypotheses suggest that the process of developmental change is in general favorably affected by the growth of cities and urban life styles across a wide range of communities and nations. But international urban research is beginning to produce other hypotheses about the associations between "urbanization" and "urbanism," and between the growth of urban areas and the process of development.

First, while Western characterizations of urban development show an association between the existence of large, dense, permanent settlements of socially heterogeneous individuals and urbanism as a way of life, in many cities elements of one occur without the other. On the one hand, urban characteristics appear outside of cities. For example, Malaysia's land development schemes, with an average population of 1,980 persons, offer a comparatively urban way of life, yet they barely approach the size of cities. On the other hand, individuals and groups may move spatially into an urbanized area and still retain nonurban values and practices. Rangoon, Burma, is large and dense, but the people who live there continue to carry on the kinship and ceremonial practices established by their rural kin.

Second, while it is true that urban development is associated with social and political development on a world-wide basis, within regions and cities there is frequently a disjunction. Under certain circumstances urbanization can slow down economic and political advances. Where the expansion in urban population outdistances growth in manufacturing and modern industry, rural migration increases urban unemployment while decreasing national food and raw material production. At the same time, the growth of cities on the basis of massive rural immigration creates tensions which, if translated into political forces, are disruptive of the existing order.

Within cities the process goes something like this: Displaced peasants seeking jobs and better conditions in the city place great strains on urban government to provide services. The municipalities are normally underfinanced. They find it difficult to cope effectively with growing service needs, or to increase their own revenues by enforcing the collection of new taxes. Instead of facing the real needs of the urban masses, they resort to symbolic satisfaction. They may construct a single lavish public housing project after razing one slum, or erect a palatial office building. Although many recent urbanites feel relatively advantaged in comparison with their rural standard of living, the greater contact of deprived city residents with more privileged modes of life can also sharpen discontent. In addition, the physical density of the city makes organizing the discontented easier than in the countryside. Tensions are also increased by the greater contact in the city between tribes and religious or social groups which, while normally antagonistic to each other, are kept apart by geography outside the city. If these forces combine to reach the required critical threshold, then urbanization without industrialization, in the absence of strong political institutions, produces heightened political instability.

Nor does urban development always promote individual modernity and satisfaction. One can certainly find many changes in values and life styles which result from the growth of cities, but these changes may threaten rather than enhance individual gratification. In America the concentration of people in urban areas has in the past tended to impose impersonality, heterogeneity, and anonymity on the individual. This produces social fragmentation and disorganization among the individuals living in the urban world. In turn, the individual's feeling of political and social impotence is increased and his political participation decreases.

Using these pictures of urbanization, urbanism, and development, many kinds of association can be seen. The customary picture suggests that urbanization produces urbanism which produces development. But observation on a compartive international basis indicates that many other combinations and permutations are possible. Take, for example, the case of Latin America. Migration to Latin-American cities frequently occurs in communal groupings called "invasions." Individuals do not move to the city alone, but with large groups from their province, or into areas to which family friends have already moved. Thus extended family ties in cities are often even stronger than among similar groups in rural areas. Also, the Latin-American migrants show less evidence of impersonality, secularism, breakdown, or irreconcilable differences between the generations than Western experiences lead us to expect. Here we have a case of urbanization without the customary forms of urbanism on the individual level.

The retention of family ties makes migration to the city easier for the individual. It decreases the insecurity the urban move might otherwise bring. Hence, there is less individual alienation. We might conclude that at this level, urbanization without urbanism is functional for development. Of course, the retention of family and rural ties also means that challenges to traditional leaders develop only very slowly in migrant communities. Personalistic leaders, displaying gifts of goods for the very poor, retain power as they did in rural regions. At this level it might be argued that urbanization without urbanism is inimical to development.

According to these alternative hypotheses, urban development viewed from a world-wide perspective seems fraught with paradox. Urban development is simultaneously linked with forces which may impede national development and forces which promote it. Residence in the city is connected both with new forms of individual politicization and with weaker participant roles for some groups. Are these findings contradictory? Should cities which deviate from the Western model be regarded as aberrations? Or should we conclude that Western concepts and theories about urban development have little relevance for explaining the results of similar spatial processes in different cultural contexts?

It is very difficult at present to draw a unified theoretical picture of international urban development which brings all the customary and alternative hypotheses together in a common and consistent framework. However, there do seem to be certain themes which result from tracing both sets of explanations. These themes revolve around the conditions under which the effects of urban development around the world may vary.

The first theme concerns the role of situational and cultural variables. Most theories about the developing city arose from the study of cities in the West. In consequence there is a need for caution in transferring concepts to nations experiencing urban development under conditions of more extensive nationwide change, and where deliberate intervention by the national government into the course of the urbanization process is more common. Two types of factors are particularly important in accounting for similarities or differences in the impact of urban development around the world. These are the so-called "situational" realities, or factors which pertain peculiarly to concrete events in a particular urban area, and the "cultural" continuities, or the values and norms which infuse the atmosphere in which urban development is taking place.

Researchers seeking for generally applicable theories of urban development often tend to downgrade the importance of situational variables, although at any given point in time the impact of urban development within and among cities and nations may be importantly affected by such realities. For example, classic thinking about the spatial organization of cities includes the theory that the modern American city takes the form of concentric rings or zones, with the wealthiest or commuter zone farthest from the center. Lagos, Nigeria, and Bombay, India, to name only two, have very different shapes. One is on an island, the other is on a peninsula. Latin-American cities show a concentration of high-value residential areas toward the center. Low-income groups and recent migrants are ranged on the outskirts, as in the medieval city. This is partly because of the impact of a traditional concentration of wealthy individuals around the inner plaza in Old and New World Spanish cities.

Another important situational factor is the presence or absence of particular leaders in cities at opportune moments. Students of the housing problem maintain that housing deficits of astonishing proportions are almost inevitable in rapidly growing urban areas, given the enormous in-migration of population and the overuse of existing housing supplies. Yet a strong leader can exercise such influence on housing programs that decisive upgrading of the quality and quantity of housing occurs. Dr. B. C. Roy's activities in promoting the reclamation of salt lakes marshland in Calcutta is an example.

The importance of who holds office, where a city is located, what the vagaries of mood are on a particular day, seems obvious when boldly stated. But we are frequently prone to forget that the impact of urban developments around the world may be radically altered by such nonrecurring events. Urban development also must be expected to change with variations in the norms which animate the culture in which urban processes act. For example, the problem of creating metropolitan governments is likely to have very different dimensions in societies where different weights are placed on identification with the locality group.

Latin-American and United States experiences with metropolitan reform are cases in point. In the United States it has proved very difficult to expand the boundaries of urban governmental jurisdictions to unite suburbs and core cities. These areas are interdependent in respect to economy, transportation, and communication, but the cititzens in each area have formed attachments within their separate localities. They oppose metropolitan consolidation or annexation out of a desire to retain small governments, close to the grass roots. Residents in the core city, with different social, racial, and economic characteristics, are viewed as a threat to the locality group.

In Latin America, however, the substitution of locality groups for kin groups and the differentiation of the city from the surrounding rural countryside has historically failed to occur. There is also less tendency to draw lines between units of government which are close to home and those which are at a greater functional distance from the individual. Instead, social formations based on coalitions of individuals and families are the common mechanisms for gaining public and private ends. On this basis alone, it may be easier to organize governments for metropolitan areas in Latin America than it is in the United States.

A related example may be drawn from the tendency of Western urbanists to favor efforts to involve citizens in local decision-making. It is argued that the development of efficient, as well as democratic, city government depends on grass-roots participation. Apart from the fact that wide participation may not be functional in expected ways in societies whose development problems still are national in scope, historical traditions also may make it inappropriate. In Africa, the building of urban participation often has been designed to further the control of those groups opposed to incumbent national leaders. Coming after, rather than before nationhood, the stimulation of grass-roots participation in this guise rankles. It should be remembered, however, that in nineteenth-century Britain, too, fears of the consequences of urban participation were expressed. The massed working-class populations were seen as "monster clots of humanity" whose uprising would unleash forces to shatter the existing social order. In India, under British rule, cities were identified both as the cores of the imperialism of colonial administrators and as seedbeds of revolution. Since independence, this heritage has persisted in a potent anti-urban ideology, also shared by the British well into the twentieth century. In India this perspective is today fed by the burdens urbanization imposes on the economy. Traditions which suggest cities are seedbeds for rebellion or threats to national progress are hardly conducive to belief in the virtues of local autonomy.

An adequate explanatory or predictive scheme for determining the effectiveness of metropolitan government plans, the applicability of mechanisms fostering grass-roots participation, or the nature of the impact of urban development around the world must take into account a wide range of both cultural and situational variables. The spread of levels of urbanization common in the West should not be expected to wipe out, or be untouched by, basic differences in values and events. That different urban styles exist is not surprising. Indeed, uniformity would be more unexpected, given the difference in contexts in which urbanization is experienced.

We must be careful to avoid distortion in discussing the inapplicability of Western theories and methods of analysis to the study of international urban development, however. In the first place, it is useful to recall that there is diversity in urbanism among the industrialized Western countries as well as between Western and non-Western nations. American urbanism has been far more associated with ethnic cleavages than that of Western Europe. British urbanism has been more characterized by the promotion of an industrial working class, and ideologically-oriented political expression on a group rather than an individual basis, than has urbanism in the American context.

There are also parallels between Western and non-Western urban processes, both at the theoretical level and with regard to problems faced by urban administrators around the world. Along with urbanization come new requirements for housing, schools, and sanitation. The circumstances in the developing world may be more acute because of the lack of financial means to substantially alleviate service crises and the shortage of qualified individuals to fill key positions. But, in the United States, as well as in India and Mexico, the proportion of existing city plans for services and areal development is very high in relation to their execution. The world over, slum clearance runs counter to laws regulating land tenure, property rights, and residential immobility. Therefore renewal programs can be implemented only after long periods of negotiations and persuasion.

The Development Board in Lagos, Nigeria, adopted a slum clearance program because they felt that the appearance of the city did not befit a national capital. Although the project was eventually implemented, the conflict unfolded much as it might have in an American city. Residents resisted the move. The lack of alternative housing created slums adjoining the cleared area. Much of the site remained undeveloped because of a lack of private funds to finance new construction. An interesting footnote to the slum clearance problem is that urban renewal in developing nations seems to require a keen understanding of the human necessities of urban life and thus demands personnel with skills only now beginning to be developed even in the United States. In Lagos, many people dispossessed by renewal were rehoused in a project on the city's outskirts. Most were miserable there, despite improved physical conditions, because of the disruption of family and social ties. The community life for which Lagos neighborhoods were distinguished was never restored in the new site.

A balanced view of the impact of culture and situation leads to the conclusion that some findings do have cross-cultural applicability while other components of Western urban theory are inadequate for the study of cities on an international basis. The separation of theories about urbanization, urbanism, and urban development which relate to Western culture or industrial development from those which are intrinsic to urban settlement anywhere will be valuable in understanding both international phenomena and urban problems within the United States as well.

A second theme which may help us to explain the presence of both positive and negative associations between urban development and political, economic, and social development has to do with the relative levels and rates of urbanization present in different countries. One interpretation of the differential impact of urban development is that it results from the presence of different levels of maturity in urban and other sectors. It may be true that the level of national development tends to rise as increases in the level of urbanization occur. But progress in particular cases still can depend on balance or imbalance with different institutional and social spheres. Urbanization which exceeds a particular level in a country at one stage of development can create problems of coordination, anomie, and service inadequacies which exceed the "solution capacity" of the political system. Urbanization below a particular level in a country at this same stage of development may be insufficient to support industrialization, stimulate the agricultural sector, or provide a locus for complex nation-oriented politics.

Different rates of urban and social or political development may also account for the seemingly contradictory impact of urban development viewed from an international perspective. For example, if the rate of urban growth in a nation is 2 per cent one year and jumps to 25 per cent the next year, this might generate obstacles to further growth in other respects. The same could be true for the growth rates of particular cities. It

247

is interesting that one city which many observers feel has made considerable progress in solving its social and service problems, Stockholm, Sweden, has had a great deal of time to meet objective needs and correct mistakes in comparison with other cities around the world. From 1950 to 1960, Stockholm's population increased 15 per cent, compared with rates of 82 per cent in Karachi, Pakistan, 104 per cent in Davao, the Philippines, or 206 per cent in Lagos, Nigeria. The growth of particular kinds of institutions or capabilities may occur more successfully in urban systems where growth is slow, major problems are widely spaced over time, and fewer needs must be filled simultaneously.

These examples illustrate that if a given level or rate of change in urban concentration is correlated with development, an increase in urban concentration does not necessarily promote further progress. Much has been written about the desirability of regulating the flow of migrants to cities in developing areas, on the theory that urban growth often unhinges economic and social development. A few countries, such as Rhodesia and the Republic of South Africa, apply restrictive legislation to migration to cities which reach a set size. The term "over-urbanized" has been coined to suggest "excessive" urbanization. The use of such glib words should not obscure the likelihood that there is no international magic number which describes an urbanization level or growth rate which promotes development. The impact of the proliferation and growth of cities is beneficial or harmful according to its balance or imbalance with other social and political institutions. This conclusion suggests the need for a dynamic picture of urban development to interpret the evolution of cities and the societies in which they are situated. A picture of this type might show what combinations are most conducive to different changes in society or the economy.

At the present state of our knowledge we cannot be certain which of the foregoing accounts of the impact of international urban development is more correct. No student of international urbanization can fail to observe that with the advent of city concentration various changes have taken place in a two-way process of disintegration and reintegration. Viewing urban life from a cross-national perspective suggests that no single pattern is "normal" or "deviant." Like other social processes, the impact of urban development is neither consistent nor tensionless. The job of international urban researchers is to specify the similarities and differences as the beginning of understanding why the forms and effects of urban development do vary.

### Note

I am grateful to James F. Guyot of UCLA for a discussion of the urbanization of the countryside in Malaysia.

# 20 SQUATTER SETTLEMENTS IN DEVELOPING COUNTRIES

John F. C. Turner

In many of the rapidly urbanizing countries of the world, urban squatters settle more land and build more dwellings than the private or commercial and the public sector combined. As the clandestine developers' or squatters' action is illegal, the settlements they establish do not officially exist. Without undermining the frail structure of government, the authorities cannot deal with squatters as they would deal with law-abiding citizens; so, if the squatters cannot be removed, they are generally ignored. This policy becomes less and less realistic, however, as the proportion of squatters grows. Squatter settlements in most major cities in Latin America, for instance, are growing at over 10 per cent per year—doubling every five years or so—and they already constitute more than a quarter of these cities' areas and populations. In some cities of Africa and the Middle East, over one-half of the city populations are squatters.

If rapid social and economic development in an urbanizing world depend on the orderly participation of the mass of the people in the processes of planned development, as I assume we all believe, then it is difficult to exaggerate the importance of this squatter problem. Its solution may well be vital for all if, as more and more believe, the main threat to world peace is the widening gap between the rich nations and the poor. It is the so-called Third World—the poor two-thirds of mankind—that is urbanizing most rapidly and whose immediate future depends on the very rapid dissemination of urban and industrial skills.

From these premises it follows, without further argument, that for governments to ignore large and rapidly growing sectors of the city populations in the developing world is to invite chaos and perpetuate poverty.

There can be no doubt that the overwhelming priority among contemporary urban problems is to find the answer to the questions posed by squatters in the urbanizing countries. I have no doubt, personally, that the answers to these questions are clear from the way in which they are stated: that if we read what the people have already written into contemporary urban history and if we listen to what they say—often literally—we will learn the answers. Not only the answers to the problems in Ankara, Kinshasa, Calcutta, and Caracas but also some very significant clues, if not the whole answer, to the problems of Los Angeles and Detroit.

The heart of the problem is in the reading or interpretation of the facts of urban growth and squatter settlement. In view of the immense scale and rate of anarchic squatter or autonomous urban settlement, it is astonishing that so little is known. Both the general news reader and the specialist obtain the same image from the bulk of the reports on urban conditions and problems in the Third World. The following quotations and the images they conjure up are typical:

> All around Lima lie these dreadful slums, the notorious *barriadas* of Peru, in which 250,000 people live like gutter-creatures in the dirt . . . slum[s] so bestial, so filthy, so congested, so empty of light, fun, color, health, or comfort, so littered with excrement and garbage, so swarming with barefoot children, so reeking of pitiful squalor that just the breath of it makes you retch.[1]

That is reporter James Morris' view of the squatter settlements of Lima; sociologist Sam Schulman's view of the equivalent in Bogota is much the same:

> It is the rudest kind of slum, clustering like a dirty beehive around the edges of any principal city in Latin America. In the past two decades poor rural people have flocked to the cities, found no opportunities but stayed on in urban fringe shantytowns squatting squalidly on

the land. . . . Living almost like animals, the *tugurio's* residents are overwhelmed by animality. Religion, social control, education, domestic life are warped and disfigured.[2]

William B. Walsh, a doctor, caps these with his description of a squatter settlement in a provincial capital in northern Peru:

In this enormous slum lived some 15,000 people, many of whom had come down from the mountains, lured by communist agitators. Why starve on the farms, the agitators asked, when well-paid jobs, good food, housing, and education were waiting in Trujillo? This technique for spreading chaos and unrest has brought as many as 3,000 farmers and their families to the *barriadas* in a month. Once they arrive (on a one-way ride in communist provided trucks) they are trapped in the festering slums with no money to return to their farms. . . . But there are few jobs for farmers in a city and they soon find that, as residents of a *barriada,* they are despised outcasts.[3]

Briefly, the image that we get from these reports published in the most widely distributed periodicals—the above are from *Reader's Digest, Life* magazine, and *The New York Times*—is one of extreme poverty and squalor, of disorganized masses of unwanted and unemployed rural migrants, surviving through begging, crime, and prostitution and all hovering on the brink of violence into which they are easily led by subversive demagogues. It is obvious that if squatters are like this, then they and their settlements are parasitic burdens on the economy and grave threats to development or, even, to institutional and economic survival.

From the growing number of less widely circulated studies and reports by observers who had close contact with these areas over long periods, one can obtain a completely different image. James Morris' horrifying picture of physical squalor is contradicted by William Mangin's description of the same areas (in Lima, Peru):

*Barriadas* are located on state, municipal, or church lands in or near the city. They were originally formed by some sort of an organized invasion and each has a formal association and is considered a community by the residents.

At its worst a *barriada* is a crowded, helter-skelter hodge-podge of inadequate straw houses with no water supply and no provision for sewage disposal; parts of many are like this. Most do have a rough plan, and most inhabitants convert their original houses to more substantial structures as soon as they can. Construction activity, usually involving family, neighbors, and friends, is a constant feature of *barriada* life and, although water and sewage usually remain critical problems, a liveable situation is reached with respect to them. . . .

For most of the migrants the *barriada* represents a definite improvement in terms of housing and general income, and Lima represents an improvement over the semi-feudal life of the mountain Indian, *cholo,* or lower-class mestizo.

There is very little violence, prostitution, homosexuality, or gang behavior in *barriadas.* Petty thievery is endemic throughout Lima, but *barriadas* seem somewhat safer than most neighborhoods in this respect, perhaps because there is less to steal.[4]

Sam Schulman's ferocious description of a squatter community in Bogota, Colombia, is also contradicted by another report by Charles B. Turner describing another but similar settlement in the same city, one which I visited myself a short while ago:

Policarpa is alive with activity—and not the stolid, passive going about the day's labor type-of-activity one finds in Las Colinas [the settlement which Schulman refers to]—there is much building and rebuilding as the new residents apply improvements to their houses and further solidify their claim to the land. Each week it has been possible to see the results; as houses are converted from tarpaper, to wood or gadua, to more substantial and liveable structures.

The results are proof positive of the low-income family's self-reliance and ability to improve. It is not difficult to discern the sense of pride and satisfaction of the people working on their own home manifest when asked, how long it took them to build it, or what they plan to do next.[5]

When I visited Policarpa in July, 1967, three years after

Charles Turner made his observations, a large number of these provisional or semiprovisional houses had been converted into brick and concrete structures, and the highly organized community had installed a piped water supply and were laying down sewers.

William Walsh's story about the supposedly subversive squatters of Trujillo (also in Peru) is as absurdly inaccurate for that city, which I happen to know well myself, as it is of the *gecekondu* squatters of Ankara to whom Granville Sewell refers:

> Government officials and intellectuals in Turkey have frequently expressed concern that the residents of the *gecekondu* will become dangerous radicals. . . . Despite the substandard living conditions, however, several forces are operating to counter such a trend at this juncture. The migrants are principally villagers with a deep devotion to their religion and a surprisingly powerful sense of Turkish nationalism . . . . Secondly, the vast majority of the *gecekondu* residents have accomplished significant social and economic mobility in a relatively short period of time. Most "own" their homes and have adequate sources of income. As long as they can maintain their sense of social improvement, the probabilities that they will become highly dissatisfied are small.[6]

To take the diametrically opposite view to that presented by the popular press would be dangerously misleading as well, of course. There are a number of very important variables, which I cannot discuss completely here, but I will mention a few of the most important ones. The general social characteristics of those who are forced to squat in order to house themselves vary enormously, so that the physical characteristics of their settlements also vary, according to the demographic and economic contexts.

The rate at which urban populations grow is a function of the urban birth and death rates, of the over-all growth rate of the hinterlands from which the cities draw migrants, and of the distribution of the population between the terminal urban areas and the rural and the intermediate urban locations. At one extreme, that of a slow-growing, predominantly rural population, migration to the cities may be very small, adding only a little to the slowly growing—or even shrinking—native urban population. This was the usual situation in all countries before the start of the Industrial Revolution in England in the eighteenth century. It is still the situation in some countries, in Bolivia and in Haiti, for example, or in Afghanistan and Nepal. Though relatively fast by historical standards, the growth of cities in these countries, and a rapidly diminishing number of others like them, is slow by modern standards: in Bolivia, for instance, less than one-third of the population lives in towns of over two thousand inhabitants and the towns are growing at about 3.5 per cent per year. Colombia, on the other hand, a country typical of the rapidly urbanizing majority, has half of its population in urban areas already, and these areas are growing at 5.5 per cent annually. This is because of the very much greater pressures on the village and small town populations to migrate owing to the poverty and insecurity of rural life.

At the other end of the spectrum, in the countries already highly urbanized, like Uruguay (to continue with the Latin-American examples analyzed by Lowdon Wingo), the urban populations are also relatively slow-growing. In Uruguay over 80 per cent of the population is already settled in the urban areas, so that the effects of migration are proportionately small, however many of the rural population migrate. The urban areas of Uruguay are growing at a bare 2 per cent annually. A few moments' reflection will be enough to see the connection between these demographic variables and the nature of the housing demand: in the rapidly urbanizing countries a very high proportion of newcomers are young adults with families or about to establish them—a very high proportion of the expanded population is demanding separate dwellings, therefore. If the new population were all infants of existing families the demand would, of course, be entirely different until that generation became adult.

Another very significant variable, and one that correlates strongly with the demographic aspect of urbanization, is the economic situation. The degree and rate of industrialization directly affects the nature of the demand for labor and, of course, the wage levels that it can earn. The rapidly urbanizing and rapidly industrializing countries, such as Mexico and Venezuela, will give their urban populations much higher income levels than the rapidly urbanizing but less rapidly industrializing countries such as the Philippines or Brazil. This at once suggests that there will be major differences in the quality of housing demanded, even if the quantity is much the same. In Manila and in Rio de Janeiro there is a far greater demand for very cheap and relatively temporary accommodation than in either Caracas or Mexico City. The Venezuelan and Mexican wage earners have higher and more secure incomes and are, therefore, much more interested in the possession of a permanent house which they can afford to build solidly and well. The poorer and less secure counterpart in a Rio *favela* or in a *mocambo* of Bahia, on the other hand, prefers to have a cheap shack that he can abandon or move when it becomes necessary to look for a new job in a new location. This, among others that I cannot go into here, is one of the reasons why there are such enormous differences between the *favelas* of Rio and the *colonias proletarias* of Mexico City; or between the *gecekondula* of Ankara and the *bustee*s of Calcutta.

With such powerfully determinant variables, it is evidently absurd to argue about any one generalized world image of squatters and their settlements. In fact, the variations are almost as great as those to be found among normal small towns throughout the developing world. Some *barriadas* in Lima, Peru, have higher physical standards than any parts of many provincial towns in Peru, while other *barriadas* also in Lima, may well fit Sam Schulman's lurid description of a *barrio de invasion* in Bogota.

While I have not the time here to explain the whole spectrum of squatter settlement types, I can clarify the polarity implied by the quotations. There is some truth in the conventional image: there are squatter settlements that are slums by any definition of the word and whose inhabitants are, indeed, wretchedly poor. Some may even have enough spunk left to react in violent protest. But these settlements, which are usually located on central city land, contain a relatively transient population and will sooner or later disappear. At the other extreme, however, there are many squatter settlements that are socially de veloping and physically self-improving suburbs, rather than slums. These are almost always located at the edges of the urban areas. Their inhabitants are generally poor, but less poor than the central city slum dwellers; they are stable residents, *de facto* possessors of their own plots and builders of their own permanent homes. Few of these squatter suburbs will be eradicated. The vast majority are in the process of administrative as well as physical integration with the city proper. Between these slum and suburban extremes there are, of course, many mixed types —probably a majority. The very common and frequently reported hillside settlements—the majority of the *favelas* of Rio de Janeiro and the *ranchos* of Caracas, for example—usually exhibit both suburban and slum characteristics. Without careful qualification, generalization is impossible.

In spite of the remarkable variety of regional and local conditions, national policies for dealing with squatters and their settlements are extraordinarily uniform. A short while ago I was in Barcelona, Spain, where I saw the remains of brick and concrete houses demolished by the city authorities because they were clandestinely built. In the recent past the same policy has been attempted in Ankara; to some extent it was actually carried out in Venezuela under the dictator Perez Jimenez. It has been threatened and occasionally attempted in dozens of countries and in hundreds of cities. But where significant proportions of the urbanizing populations are obliged to build illegally for the lack of any alternative, these attempts have invariably failed. This failure is not at all hard to understand, in view of the scale of urban settlement and the dependence of any modern government on popular support. Less obvious, perhaps, are the reasons for the very general failures of the relocation programs and the large-scale low-cost housing projects designed to meet the popular demand. Whether one visits Barcelona or Bogota, Caracas or Karachi, one is bound to find that the major government housing efforts are dedicated to constructing large projects of minimum modern standard dwelling units. But one invariably finds that the supply is totally inadequate for the demand and that, in fact, it has had little effect on the continued existence or growth of the squatter settlements.

Even the immense and long-sustained squatter relocation program of Hong Kong—a city with a colonial government un-hampered by political discontinuity and with an exceptionally big budget—has only managed to hold the squatter areas at the level they achieved when the program began over ten years ago. The huge but unsustained—and very badly administered—"super-block" program carried out by dictator Perez Jimenez in Caracas had little or no over-all effect on squatter settlement growth; indeed, some experts have claimed it actually stimulated *rancho* growth by attracting even more migrant workers to build the immense structures. The apartments provided in them, by the way, were far too expensive for the squatters, so that the new buildings themselves became squatter settlements—and thus extremely expensive for the city. Many major and highly publicized squatter relocation projects are so badly behind in recovery of the investment made that they would be bankrupt if it were not for government support and subsidy. (I know of two where the occupants are reported to be over 80 per cent in arrears with their rents or mortgage payments.) When all the facts are known, which is quite rare, it is exceptional to find that a "low-cost" housing program is, in fact, low-cost at all. One authority has analyzed costs of government housing programs throughout Latin America; his findings show that, if the families for whom the projects are designed had to pay their actual cost, over 80 per cent of the "low-income" sector would be unable to occupy them. A little arithmetic will convince anyone that government, or even international subsidies are no answer to this problem. The quantitative problem is far too great in relation to the financial resources available. Orthodox housing policies, along the lines of the industrial countries, do not and cannot prevent squatting in preindustrial or transitional and urbanizing economies.

These failures, it is clear to me, are the natural result of gross misconceptions of the problem—of messages misread and, consequently, of goals misplaced. The policies practiced would be logical, at any rate, if the premises on which they are based were correct. I do not think I am far wrong in saying that the popular press image of the squatter and low-income settlement problem is the same as that of almost all governments, irrespective of the demographic and economic situations of their countries. The uniformity of their actions certainly supports this assertion, as the uniformity of their failures suggests that something is universally wrong with their criteria or methods.

If it is assumed that squatter settlements are the direct product of rural migrants who cannot support themselves in the city, the logical policy is to tackle the problem at the source and prevent the migration. This has been tried—very seriously in the cases of South Africa and mainland China, which both have exceptionally powerful administrations—but with no significant success. Two facts, applicable in most regions, are enough to explain the general futility of otherwise seemingly sensible migration control policies: the very rapid rate of population growth in densely populated rural areas on the one hand and, on the other, the reduction of the demand for agricultural labor as productivity rises with modernization. In other words, there are far too many people on the land already—in most countries of Asia, Latin America, North Africa, and Middle East—and the rapid increase of food production demands fewer, not more, people on the land. In some areas, especially in Africa, agricultural colonization and stabilization may reduce the pressures on the cities, but even so, it will be only a temporary relief. In other areas, in Colombia, for instance, increased food production requires a rapid decrease of the rural population and, therefore, an even higher rate of urbanization. One authority estimates that 70 per cent of Colombia's rural population should be relocated in the cities in order to achieve the rate of agricultural production necessary to feed the whole population properly.

Whether convinced by such facts or not, most governments accept the urbanization of the population as inevitable and virtually uncontrollable. But, because they regard squatters and their settlements as undesirable, all their efforts are oriented to the eradication of the settlements and relocation of squatters in modern, though necessarily minimum standard, dwellings. These relocation programs are often accompanied by community development, educational, or employment programs—few administrators these days are naive enough to suppose that poverty and ignorance will be sloughed off with the slums themselves. A government that is aware of its inability to reverse the process of urbanization, and that cannot ignore the problems created and survive politically, will do what it can to improve environmental conditions and to speed the process of urban acculturation through education.

These apparently irreproachable aims do not appear to be so reasonable if a different view is taken of the nature of the problem. If the picture given by observers like Mangin, Sewell, and Turner is correct, then the attempt to substitute government action for popular initiative and the presumption of the would-be "educators" and "community developers" seems absurd and doomed to failure. In the ground-level observers'' view, which I share, squatter settlements are not barriers to urban acculturation so much as vehicles for upward socioeconomic mobility; many —in some areas the great majority—of the squatters are not ignorant and newly arrived peasants but able and active, though poor, working-class families. Most of the adults are from provinces, but this is true of the majority of the adult population of all rapidly growing cities and, though they are migrants, most squatters arrived years before and are quite well adapted to urban life—even if they weren't already adapted before they arrived. In this view, this sector is no more—and no less—in need of education than any other, and in general it has far less need of "community development." When the external conditions are right, working-class groups intent on building up their own homes and local environments develop admirable communities without any need for outside interference. As a middle-class, college-educated professional I feel ridiculous trying to tell these people how to organize their own lives in their own environment. I have tried it and have learned far more from them than I could ever teach. I have served such communities, I hope usefully, as an architect: I have provided better designs for schools and houses than they could make for themselves—but that is my job and, as a professional, my relationship to a village or urban community is no different than it is to a wealthy individual client or an institution. By the same token, a government agency that shares this view will not attempt to substitute its own scarce resources for those which the people possess in relative abundance. To build for people who are so eminently capable of building for themselves, or to try to organize people whose spontaneous organization is often strong enough to defy armed force successfully, deserves to fail and is bound to do so if attempted on a large scale.

A rational policy based on the alternative interpretation of the problem would be to provide the people with resources that the government possesses or controls and the people do not. The first item, of course, is land for those who want to build—and who can and who often will build on land taken by force if no acceptable alternative is offered. The second and third items are technical assistance—to ensure that the land is properly subdivided and the buildings properly designed and built—together with the essential community facilities or services: public transportation, retail shops, schools, and so on. Public utilities may or may not be very important, according to the locality and the resources and demands of the local community. In a great majority of cases, publicly or commercially built structures are not needed initially; they are far too expensive. Most low-income families who feel secure enough to settle down in one place and invest their savings in a home of their own much prefer to build it in their own time and way; they can build far more cheaply, and by building in stages they can avoid the insecurity of indebtedness and mortgages, as well as the very considerable extra cost of interest on long-term credits.

Tentatively, and on a small scale, programs of this kind have been experimented with; the only projects I have seen that are successful by the criteria mentioned earlier are those where the sponsors limited their contribution to land or land titles, technical assistance, and credit for building materials and skilled labor. The projects of this kind that I have followed or seen in Latin America have stimulated a far greater ratio of investment by the beneficiaries than that which the government made or could make on a significantly large scale; the administrative work and subsequent problems were negligible in comparison with the usual government-built housing projects; and the environments themselves are far less dreary than the usual institutional project and, when complete, will be typically varied and "natural" communities both socially and aesthetically. Projects such as these are the expression of a very unorthodox policy alternative: one which seeks neither to suppress what the people do on their own initiative nor to supplant it, but rather one which supports popular initiative and is based on the huge resource which that initiative represents.

This view challenges many current assumptions. Unquestioning faith in the economy and social necessity of vast organizations —whether private capitalist or state socialist—has blinded us to the evidence of the potential and value of local direct action, at the scale where democracy has a very real meaning. Those of us who have come to depend on a smooth and highly institutionalized modern life are a bit afraid of recognizing the values and reality of a more independent life, and it seems that we are most unwilling to acknowledge its products or potential. It is this attitude, I am sure, this faith in complex organization and its gadgetry, that has led the great powers to make their appalling and dangerous mistakes in the Near and Far East—to forget that achievement is a product of people's will. The attempt to substitute bureaucratic or technological machinery for personal will and initiative is futile and, in the longer run, suicidal. Poor countries, those of the Third World, have one immense advantage over their wealthier neighbors: they cannot afford to make the same mistakes. By facing the fact that they cannot substitute machines for men, the developing countries may well show us how to use our own sophisticated tools in humane and genuinely constructive ways. Their only chance for development and survival, after all, is to use the few tools that they can afford to stimulate and support the initiative of the mass of the common people.

## Notes

1. James Morris, *Cities* (New York: Harcourt, Brace & World, 1964).
2. Sam Schulman, "Latin American Shantytown," *New York Times Magazine* (January 16, 1966).

3. William B. Walsh, "Yanqui Come Back!" *Reader's Digest* (December 1966).

4. William P. Mangin, "Mental Health and Migration to Cities: A Peruvian Case," in Dwight B. Heath and Richard N. Adams, eds., *Contemporary Cultures and Societies of Latin America* (New York: Random House, 1965).

5. Charles B. Turner, *Squatter Settlements in Bogotá* (Bogotá: Centro Interamericano de Vivienda y Planeamiento, 1964).

6. Granville Sewell, "Squatter Settlements in Turkey: Analysis of a Social, Political, and Economic Problem" (Unpublished Ph.D. dissertation, Massachusetts Institute of Technology, 1964).

# 21 HOW TO VIEW A CITY

## Martin Meyerson

If man is to improve the art of living in cities, he must become more knowledgeable about himself and about his surroundings. It is impossible to have richer experiences of urban living without understanding what has been, what is, and most importantly, what might take place in urban areas. I have chosen to discuss how to view a city (and when I say a city, I mean, of course, an urban area), because I believe that we literally do not see our cities; we are unconsciously familiar with them in a way which blinds us as much to their realities as to their future potentialities. Before we can introduce more grace, or more order, or more tolerance, or more productivity, or more humanity, or more diversity, or more beauty in urban living, we must bring into conscious awareness whether it is these or other qualities that we seek, and we must examine those urban satisfactions which now exist and might be increased or which might be newly created.

There are many ways, of course, of looking at a city. Throughout history, the different observers have regarded the city in different lights—some have seen it as a place of authority, others as a marketplace; some have seen it as a place of iniquity, others as the very foundation of man's civilization; some have seen it as the disrupter of human values and stability, others as the place where spiritual and other human values are perpetuated; some have seen it as a collection of the disorganized and the disruptive, others as a place where individuality can flourish. Some have seen the city as a place of chaos, others of vitality. These days some people in all the industrial countries of the world look at the city as dying—choking on traffic, starving from lack of fiscal nourishment, bleeding through loss to the outskirts of population and economic enterprises, and worn out with obsolescence. And others, including myself, point to the unfailing magnetic power of the city, which continues to draw people away from the farms and from the small towns, not only in industrial societies but in nonindustrial ones as well. Depending on one's philosophic bent, or political viewpoint, or temperament, or interests, one can find evidence of any or all of the characteristics I have mentioned in the large city today.

How then shall I ask that we view the city if there are so very many lenses which could be used in the examination? What I ask is that we view the city with understanding, with appreciation, with expanding expectations and aspirations, and with a sense of commitment and a sense of responsibility. We are now at the point where we can properly ask more and more of our cities; by seeking greater satisfactions in the quality of civic services, in the kind and variety of cultural opportunities, in the modernization of the economic base, in the attractiveness of the physical environment, in the convenience of the transportation network, and even, and perhaps most importantly, in the harmonious relationships in the various groups, ethnic and otherwise, among the citizenry—a problem incidentally which holds throughout the world and not just in America. We are now able to achieve these satisfactions, if we choose to do so. I shall be talking primarily about America, but much of what I say will be applicable to other countries as well.

Despite the frequent disparaging attacks on the young, we have never had a younger generation more knowledgeable or concerned about its environment. There has been no magical intervention in the arrangement or distribution of the genes to bring about such a change. One thing that has happened in America is that more and more young people expect to go to universities and colleges and actually do go. The expansion of opportunities for higher education, and not only for children in moderate- or high-income families, but for others as well, has been a part of the American genius for creating an open and expanding society. We are almost at the point where, in a majority of American families, one or more members will have gone on to higher education for one or more years.

Furthermore, recent research in education has shown the close relationship between expectations and results. Teachers of culturally deprived children who were told their pupils had outstanding abilities expected more of the pupils, encouraged them to a greater degree, and got greater response, in contrast to teachers who had not been told the youngsters had ability, and who therefore had lower expectations of their students. The level of educated citizenry we are achieving and the level of expectations about these citizens which our public and our private leaders are acquiring will have the greatest influence on the kind of cities we shall have in the future.

President John F. Kennedy, in his 1961 inaugural speech, said that we should ask not what our nation can do for us, but what we can do for it. I am suggesting that we ask more of our cities, and in asking more of our cities, we primarily ask more of ourselves. Each one of us, in what we consider to be our private lives, makes decisions which affect the well-being of the community: these decisions may be small, as in the case of the individual homeowner who cares enough to maintain his property, or the shopkeeper who attractively displays his wares, or they may be large, as in the case of the manufacturing enterprise which expands substantially and provides more jobs for all kinds of people. The decisions may be small, as when an individual goes to a concert or a theater, or they may be large, as when individuals make up the deficits of symphony orchestras or other complex cultural activities. The decisions may be small, as when a property owner rents good accommodations to a black or other minority family, or they may be large, as when a consortium of philanthropic or other organizations constructs good new rental facilities for low-income people. It is the spectrum of small and of large decisions which determines whether a community offers greater or lesser satisfactions.

And, of course, there are many civic governmental decisions, which all of us to some extent can influence, if we care to let our views be known. For example, most street furniture—lampposts, traffic signals, hydrants, street signs, benches, wastebaskets, fences, and the like—are ugly. They are constantly being replaced. If enough people cared, they could be better lettered, better designed, better placed, with very little extra resources or effort and with much more aesthetic satisfaction. Urban renewal, on the other hand, is a very complicated matter as well as costly. It takes a great deal of organizational skill and involves national and state levels of government, as well as local—but substantial progress can be accomplished, provided the people, in working together with officials, have shared goals. But we are often so familiar with our cities that we are blind, I suggest, to their realities as well as their potentialities.

No one teaches us how to get around a city and we generally are inarticulate about what experience has taught us. But even though we are not formally instructed, and even though we are inarticulate about what we know, we all find our way to the urban functions we feel we need.

We find our way through the city much as we do through a supermarket. No one teaches us; we learn by experience. We are not articulate about the patterns, but after we get used to shopping in a supermarket, we expect the milk and dairy products not to be in the middle aisles but on a side wall, and fresh tomatoes are not side by side with canned tomatoes. The foods are not grouped by degree of expensiveness; all items, for example, costing more than $1 on one aisle. Nor are the items together so that they form menus—a breakfast menu here, a dinner menu elsewhere. If you are taken back by these observations, it is not because the physical arrangements of a market are so orderly that they could not be different. I am pointing out how we accept these physical arrangements without question, without articulation. We have learned how to read the message of the market, written in large print. When we get down to the fine print, to finding out where the chutney is shelved, we have to hunt. So, too, we have learned about buildings; when we go into a new office building, we usually can find the elevator and we don't look for the newspaper kiosk on the fifth floor.

We can also read at least the chapter headings of a city —we all know, for example, that city hall is not apt to be on the outskirts of the city. We know that the affluent rarely live next to the railroad tracks. We know that they do live near each other and they generally enjoy as high a level of municipal services as is provided publicly. We know that economic reasons encourage the specialization of economic enterprises, that we rarely find a lumberyard in the very center of downtown. Thus we all have a rough, unarticulated guide to American cities and cities elsewhere, that we get through experience. When we go to a strange city, we know roughly what to expect. We know that if there is a concert hall it is more likely to be near the center than any other place.

Now, if those of us who are accustomed to cities get the message of the city written large, most of us—even the better educated among us—skim lightly through the contents, seizing on a phrase here and there, and usually skipping the footnotes. And we may realize that there may be an Italian ethnic district in one area, a Polish one in another. And most of us may realize that there are certain kinds of industrial districts, and wholesale districts, and so on—but few of us are conscious that costume jewelry wholesalers, for example, huddle together in a few urban areas, or that dentists are rarely on the first floor, or that although every city does not necessarily have a biscuit factory, it often has a bakery where bread is made because fresh bread does not travel far.

Now, there's much that we must learn when individuals of a racial or ethnic group live in close proximity to each other —is this because of self-choice or because they experience difficulty, social or otherwise, in finding housing elsewhere? Are there few frame buildings in a city such as Philadelphia because wooden construction is not permitted by the building code, or because there is a long history of stone and brick construction and the people of Philadelphia have come to prefer buildings made of such materials? Are the spacious front lawns in some neighborhoods there by law, or by what are considered the mores of quality building? So far I have suggested that city dwellers acquire from experience rough guides to the character of cities. We are not, however, even in an inarticulate fashion, usually aware of the nuances and the subtleties of the character of the city.

I have used examples from the physical environment of the city—or examples of the physical manifestations of urban functions—and I have done so for two reasons. One, it is my con-tention that we do not see even what is most visible to us because we have not consciously sought understanding and appreciation of urban patterns. And two, we have grossly un-developed urban tastes. We have not asked enough of the urban environment, we have not committed ourselves suffi-ciently to it, to improve it significantly. If we have accepted air and water pollution, overcrowded schools with underpaid teachers, undistinguished and often ugly public architecture, acres or even miles of dilapidated housing and other obsolescent buildings, and an inadequate transportation system, as some of the unpleasant but inevitable by-products of urban living, then the blame is ours.

And perhaps all of us need some help, and perhaps the poor in particular need help, in taking advantage of present munici-pal services. For example, volunteers, and often students, are beginning to teach the old, the new city resident, the poor, the minority group member how to benefit from present mu-nicipal offerings—libraries, museums, mental health services, garbage collection, housing inspection, parks and playgrounds, child guidance services, legal aid. As the American nation be-comes more and more dedicated and devoted to services, rather than to goods, a gap between professionals and potential service users may widen unless there are some means of communicating and mediating between the one and the other.

I do not think it matters by what technique the residents of the city are encouraged to take advantage of the services and the facilities that only a city can offer. And I do not believe that it is just the poor or the uneducated who underconsume in city services. But if there is a genuine expansion of urban services and a linkage of the services to the consumers, if more people are taught about their community (consciously but not necessarily formally) instead of just experiencing a partial version of it, then important results should occur. Firstly, the life of the ordinary citizen would be enriched. Secondly, the standard of community service would be raised as more and more take advantage of the opportunities newly made open to them.

I have suggested that one method of raising civic aspirations

—of eliminating what I call the poverty of desire—is to encourage underconsumers of urban services to assert a dignified and increasingly knowing interest or claim to these services. Another method of raising civic aspirations will come, I hope, from the increasing educational attainment of the population. As I indicated, the American population will have the highest level of formal education that the world has ever seen, and I hope this will be true qualitatively as well as quantitatively. Ought we not to begin thinking more and more of the richness of cultural opportunities and continuing education programs which could be offered? Touring companies in the arts have begun to grace our smaller cities for short stands. Might we not aspire to more, for example, to regional resident companies in experimental opera, and in theater and film, in television and radio, in new dance and music, and in entirely new art forms?

One group of creative people helps attract and sustain another; one intelligent, knowledgeable audience is a base from which to draw another; stimulus provokes stimulus; competence encourages more competence. If we help accelerate the tastes and aspirations of an increasingly educated population, and if we encourage underconsumers to greater use and enjoyment of the urban community, I foresee a magnificent cultural expansion in American urban areas. What I fail to see is how we could have a true cultural expansion without a deliberate, persistent program of improving the physical environment of the American urban community—the large ones in particular, but the small ones as well. The urban aesthetic is one of the least developed of urban tastes; the practice of its art is one which could produce the greatest satisfaction. Unfortunately, men of influence tend frequently to be men who do not have highly developed and visual urban tastes.

No one can escape, however, the splendor or the squalor of the cityscape except through inattention. Everyone is a captive audience; this makes the urban design challenge even more demanding and more rewarding. By urban aesthetic I mean, of course, the beauty of buildings and their association one with another, the majesty of vistas, the use of ornaments such as fountains and sculpture, the appreciation of natural resources such as lakes and rivers, the planting of trees, and the generous provision of parks and open spaces. I also mean attention to detail, such as in the lettering of signs, or the type of street pavement. Some of the urban aesthetic should be grand and some should be intimate, some should preserve the historic and some prod the innovative. But the urban aesthetic is more than visual—it has to do with the quality of air and of water, the reduction of odors, the moderation of noise. It also has to do with urban proportions, the relationship of one function to another: highways obviously should be provided at their terminus with adequate parking, but highways should not be so located that they deprive people of enjoying a park or a body of water.

Economists speak of discretionary income—that income that an individual or family has left after basic maintenance costs, to spend as he or it wishes. Discretionary income in America has been rising and an increased proportion of family and individual income is going to services rather than to things. I suggest that there is also discretionary location—many highly educated people can locate where they wish, for they have many employment possibilities open to them. They locate, then, where they feel that life will be most productive and most pleasing. More and more they judge the urban areas in which they might locate according to the educational, health, recreational, cultural, and environmental satisfactions. I predict that the quality of aesthetic satisfaction will be as important an element in environmental satisfaction as any other single element; I further predict that there will be a limit to the flourishing of the other urban arts unless they are undertaken in concert with the urban aesthetic.

We now have the national wealth and the talent to make our urban areas the epitome of beauty and social and economic and material well-being. The revitalization of urban life through the expansion of community services will require considerable rethinking of our current urban programs. It will require a large investment. In the next ten years, the Gross National Product in America should attain a level of more than a trillion dollars annually, an increase of perhaps a third of a trillion dollars over the present. Are we prepared to devote a third to a half of that increase to completely change the quality of life in our urban environment, and to do so for all people, black and white, the poor and the well off? To do so we must have the will, the understanding, the appreciation, the commitment to urban life in the particular communities in which we live. In our choices of governmental officials and support of government programs, in our individual daily business, professional, consumer decisions, in our determined efforts to become knowledgeable about ourselves and our communities, and to keep our levels of aspirations high, we should remember these words of the great art critic, the great art historian, Bernard Berenson: "All of the arts—poetry, music, ritual, the visible arts, the theatre—must singly and together create the most comprehensive art of all, a humanized society, and its masterpiece, free men."

And that precisely is what we must achieve in the cities, the urban areas of the United States and of all of the countries of the world, and I am convinced that in this, the last third of the twentieth century, no goal will be more significant for all the people of the world and that we shall succeed.

# 22 BRITISH AND FRENCH URBAN GROWTH STRATEGIES

Lloyd Rodwin

Two problems have shaped national strategies for urban and regional development. One is the lag between growing and depressed areas. The other is the effort to deal with the problems of the metropolis and decentralization. Britain and France have wrestled with both problems for the past generation. They have developed some comparable urban development strategies for lagging regions, but they are poles apart on general urban strategy. Top British planners consider their big cities too big and want to slow down and change the form of their growth. French planners feel that they don't have nearly enough large cities and that only such cities could offset the attractions of Paris. I propose in this chapter to examine the way these policies evolved in Britain and France and the kinds of problems encountered in trying to make them work.[1]

## THE BRITISH EXPERIENCE

The main elements of the British policy first became visible just before and after World War II. Two major nineteenth-century movements of reform converged during this period. One dealt with the employment conditions of the working class and gradually evolved into a more general preoccupation, among other things, with the hazards of modern industrial society, including "structural unemployment" in "depressed areas." The other concerned itself with the condition of the city and led to a persistent agitation for a healthy, efficient, and attractive physical environment.

These two reformist traditions were a response to the changes in the economy and cities of Great Britain during the past century and a half. The population had moved off the land into cities and then steadily into the larger metropolitan areas. In the process the South emerged as the principal center for trade, finance, and light industry and for political and economic decisions, and the North became the center for resource-oriented heavy industries. All of the regions enjoyed growth throughout the nineteenth century, although the South grew most rapidly. The new thriving activities were mobile enterprises—market, service, and supply-oriented. One consequence was a relative and absolute decline in some regions and an even greater concentration of new jobs in the South. By the turn of the century, there were regions in the North which had experienced a loss of population. In still others the basic trends in the relative rates of development were visible even earlier. In 1801, London and the Home Counties had a little more than one in every six persons in Great Britain. In 1861 it was one in five, in 1901 a little more than one in five, and in 1961 it was one in four.[2]

During the 1930's there was exceptional unemployment in the North. Migration soared, population growth seemed to be coming to an end, and prospects for the future looked bleak. There was widespread concern that unless something drastic was done, the South would continue to grow at the expense of the North for the rest of the twentieth century. It was these prospects which led successive governments since the 1940's to commit themselves to an explicit policy of assisting the North.

At the core of this policy were three basic aims. One was to curb the rate of growth of population and of employment in London and the southern counties. A second was to increase the scale of economic activity in the hinterland, referred to here generally as the North of Great Britain. The third was to change the form of growth of the largest metropolitan regions, particularly of London and the southern counties. (Military security was another important aim at the outset but it

soon dwindled into insignificance due to technological obsolescence.) Presumably the efforts, if successful, would reduce
the diseconomies of the congested metropolis and the unemployment and outmigration from the hinterland. They would
also spur national growth as well as growth in the less prosperous regions and rationalize the form of growth wherever it
did occur.

There were many tools to carry out this policy. One set,
designed largely by urban planners, aimed to limit the size
of metropolitan areas, especially London, by reducing the densities of inner London, by surrounding the edges of London
with a sacrosanct greenbelt, and by building self-contained new
towns in the outer areas beyond the greenbelt and in the more
distant hinterland regions. Another set, designed largely by
economic planners, restricted the building of industrial plants
in the London region while providing subsidies and other
forms of assistance for firms willing to locate their plants in
the outer areas of the South and the more distant development
districts.

Since 1960, some tools have been refined and new ones added.
In 1963, the Conservative government established the Location
of Offices Bureau, whose purpose was to encourage the shift
of office activities from congested central London to suitable
centers elsewhere. In addition, the government agreed to build
a new generation of new towns (the first fifteen were started
between 1946 and 1952), largely to meet the growth and overspill requirements of previously neglected metropolitan areas.
This policy was even more vigorously pursued by the subsequent Labor government.

At the same time the government recognized the need to
provide more attractive incentives for firms locating in the
development districts. The bulk of the assistance after World
War II took the form of renting government factories for approximately half of the economic rents and the provision of
loans at market rates of interest. In 1963 the system was
changed to provide depreciation allowances and standard grants
covering part of the cost of building and machinery. The net
effect was to reduce the average capital outlay of firms in the

development districts to 75 per cent of capital costs compared to 87 per cent in the rest of Britain and to reduce a firm's outlay to 50 per cent of expenditures for the first year compared to 80 per cent elsewhere.[3] This did not absorb all of the risks or costs. Firms contemplating such a shift had to count on higher settling costs during the first three or four years; and if there were insufficient profits during the first year to provide a substantial tax offset, the easing of cash position was of limited value.[4]

When the Labor government came to power in 1964, it took steps to control the growth of offices in London. It also established a Department of Economic Affairs and started work on a national economic plan. It also decided to experiment with coordinative Regional Economic Planning Boards staffed by senior civil servants representing the main government departments in the region. To represent the views of local authorities, universities, industry and commerce, trade unions, and other significant regional interests, the government also set up advisory Regional Economic Planning Councils.

The tenor of the new ideas was reflected in several regional reports published by the government between 1962 and 1965. Most of the reports were curtain raisers, staff proposals cleared with key local figures and submitted largely for discussion. In practice, however, because of the quest for fresh ideas, they proved influential in providing general guidance for local and national development decisions. The five-year county development plans, an important innovation in 1947, did not quite serve these needs. They were wrong in their assumptions about population growth, car ownership, and social standards; and in other ways they were out of date or lacked range and perspective. The new regional reports produced by Whitehall tried to correct these limitations, and so do a whole spate of regional studies produced since 1965 by the regional councils.

For example, the reports on the South East[5] proposed new growth strategies for the South East and for London's development—based on conservative assumptions of natural increase and no migration. One proposal by the South East Economic Planning Council favored new city regions connected to the capital by "growth sectors" along good roads or rail lines. These city regions were to be spaced beyond the greenbelt—about 60 to 80 miles from London—largely because of the hostility of the neighboring counties. Others favored concentrated linear growth through the greenbelt on the grounds that urban centers should capitalize on the advantages of accessibility and investments in transportation.

In the less prosperous regions, the reports urged the encouragement of growing points as one of the basic means of modernizing and decentralizing the metropolis, not the general development of poorer regions or the eradication of specific pockets of unemployment. There was widespread recognition that it was easier politically and more effective technically to concentrate effort on a few promising areas within the region. Here one could concentrate the bulk of economic and technical assistance, provide the infrastructure and other facilities essential for expanding and diversifying economic activities, and attract many migrants who otherwise would move to the South and many able individuals from other areas who were seeking fresh opportunities in a different environment.

It is still a little early to gauge the influence of the recent legislation and of the new ideas. There are some obvious problems. All of the regional studies and proposals have been conducted almost independently of the national economic plan. That, perhaps, is not too unfortunate, since the national plan is now being completely overhauled. The studies also have serious gaps. Economic data for the regions are poor. There is no system of regional accounts. Interregional relationships are generally slighted. The social studies are even more primitive and general. Serious as these problems are, the likelihood is that they will be dealt with eventually.

As for the policy on industrial relocation, one of the main complaints has been that the free depreciation formula was not as effective as intended because "a company had to make a profit before it could make a claim."[6] The real need, it was argued, was to help new or smaller firms to shift to the development areas and to cover the higher initial outlays and development costs for new branches located at a distance from a firm's headquarters. On the other hand the survival rate for firms which did locate in the development districts has been rather favorable; and "since 1961–1964, new factory building approved for London and the South has nearly halved; in the development areas it is up by two thirds."[7]

Despite this record and the rather favorable attitude of investors to the policy of free depreciation, the Labor government decided to tighten the controls further and to modify the system of incentives both nationally and for the less prosperous regions. In July 1965, at the same time that the government introduced strong measures to curb investment and to reduce the drain on its foreign exchange, it required permits for all new building and extensions of industrial plant of more than 1,000 square feet (this was increased to 3,000 square feet in 1966) in London, the South East, the Midlands, and the Eastern region. And in 1966–1967, the government added three other measures. To promote investment it changed its policy on incentives to spur further modernization and exports and to concentrate the impact on certain industries and regions. This was to be achieved by: (1) substituting direct grants for free depreciation; (2) limiting the benefits to investments in new plant and machinery in the manufacturing and extractive industries; and (3) providing twice the normal incentives for investments in new plant and equipment in much larger development areas. A second measure was the Selective Employment Tax (SET) which taxes payrolls but exempted and subsidized labor intensive industrial employment and tourism in the development areas. The third measure tried to quicken the effects of the new incentives and to encourage labor rather than capital intensive industries. It set up in mid-1967, for a period of seven years, a system of supplementary grants—regional employment premiums—intended to reduce labor costs by about 5-10 per cent for manufacturing industry in these areas.

The changes stirred both applause and concern. Some persons urged even more help in remedying defective infrastructure, especially housing, and more favorable grants to promote growth industries and the more effective planning of integrated industrial complexes. Others were pleased because firms could now assess direct grants more effectively than beforehand. They believed the new grants would also be more helpful in tiding firms over the initial settling-in period. To be sure, there were still some doubts whether the carrot was sufficiently attractive. The investment grants were much less expensive. A leading government spokesman estimated that they would cost £300 ($840) million less than investment allowances; and the regional employment premiums are not expected to exceed £100 ($280) million.[8]

The *Times* (London) objected that "by widening the districts to large regions, the whole idea of concentrating effort on 'growth points'—points that are showing signs of individual initiative and life—is frustrated."[9] This may become a problem, but not necessarily so. The development districts were eliminated because they focused on unemployment rather than growth, and because they were dispersed and small rather than linked to a general strategy for development of a region. The new definitions of development areas make the boundaries much broader than those of the growth centers. But there is nothing in the new policy or definitions to prevent a focus on growing points while providing more choice for the entrepreneur and a more attractive policy to defend politically.

At any rate, the policies to date seem to be enjoying some success. Since July 1966, the development areas on the whole have experienced less serious unemployment effects due to the government's retrenchment measures than the South or the Midlands. This is the first time that the North has fared so well relative to the South. The relative success of the efforts has even led to concern that the policies are creating slowly declining "gray" areas between the Development areas and the more prosperous parts of the South East and the Midlands.

Another serious criticism is that the proposals for the South East, if accepted in substance, "may further tip the scales against the economic regeneration of the less fortunate regions of Britain and further strengthen the magnetism of the South East."[10] The proposals entail huge capital investments for infrastucture and services to accommodate more than six million extra people in the region by the year 2000. This area now has close to two-fifths of the population and about one-half of the increase in new jobs. The discrepancies with the rest of the country will be all the more disproportionate if favorable consideration is also given to the proposals for the Midlands.

Despite a generation of experiment with mixed results, there is almost no disposition in Britain to give up the quest; and there is still a remarkable political consensus on the goals, albeit disagreement on some of the methods. Ironically enough, the prospects of continuance of such policies appear more probable—not simply because of the agreement on goals or because the instruments are becoming more sophisticated and effective—but because quite different and equally controversial issues now loom on the horizon. For it seems likely that the extraordinary advances in transportation and communication will modify time and cost relationships between regions. There is already some evidence of this prospect in Scotland and Devon; and if and when the changes become more visible over the nation, the big debates of the future might well turn not on how to encourage development in the North, but on how to discourage it: in short, on when and where areas in the North and the South should be opened for major urban development or preserved for recreational and other purposes—areas which the metropolises in the South and the North will be eyeing hungrily as prime targets for accommodating their overflowing populations.

## THE FRENCH EXPERIENCE

Turning now to the experience of France, let me first observe that some unusual elements shaped French regional development over the century and a half before World War II. One was the remarkably gradual rate of industrial, urban, and population growth in the nineteenth and first half of the twentieth centuries, a phenomenon not at all characteristic of the other more developed economies. Another was the dominance of

Paris and the lack of any significant competitors, even on a secondary level—a pattern more often associated with under-developed economies. More characteristic of both developed and underdeveloped economies was the increasingly visible dis-parity between regions, especially between the dominant capital city and the less prosperous regions in the West, the Massif Central, and the Southwest. Conservative cultural and political traditions in the less prosperous regions extending back almost a thousand years reinforced this pattern. So did the highly cen-tralized administrative tradition which dated back at least to Louis XIV and was intensified by the administrative changes introduced by the revolution and Napoleon to consolidate central authority.

The efforts to link urban and regional planning into the national planning process in part reflected and in part were a response to these characteristics. The foundations of this regional planning had been laid in the generation following World War II. By the end of the 1950's the principal directions of the government's policies were clear. There was a genuine recognition of the need for regional planning to complement national sector planning and of the need to reshape the ad-ministrative system of the government to serve these ends. The growth of Paris was to be restrained. An elaborate incentive system was devised to induce industrial development in other regions. The less prosperous areas were to be favored. Premiums depended on the zone, the number of jobs created, and whether a new or expanding firm was involved. Zones eligible for the most assistance were areas of unemployment; the other zones were defined by the lack of industrialization, by inadequate industrialization, and by the role of the region in relation to the effort to decongest Paris.

The new aims led to a number of innovations.[11] One of the most important was the creation of 21 programming regions. Another was the creation in each region of a Regional Prefect and a Regional Administrative Conference (CAR). These or-ganizations prepared the regional portions of the national plan, recommended public investment budgets for each region, and provided progress reports each year on the execution of the plans. Still another was the creation of the Commissions on Regional Economic Development. These commissions attempted to represent the views of groups outside the government (for example, local councils, commercial and agricultural organizations, trade unions, and professional groups).

The General Directorate of Planning supervises the actual preparation of the regional portions of the plan. It furnishes the regional authorities with drafts of the national plan, the results of the manpower studies, and the regional capital investment budget—that is, a breakdown of those elements of the national public investment expenditures which were scheduled to occur in the region. In turn, the staff of the Regional Prefect, acting in behalf of these organizations, draws up a draft of the regional portion of the plan, including estimates on costs and financing arrangements and recommendations on the priorities and linkages for the investments. These reports are then forwarded to Paris. There they are arbitrated and cast in final form by the General Directorate of Planning and the relevant ministries, and ultimately the Inter-ministerial Committee for Regional Action, which in effect functions at the national level the way the Regional Administrative Conference does at the regional level.

These innovations fixed the basic administrative machinery for regional programming. What was lacking, however, were some mechanisms which would help to monitor and coordinate regional programs of the various ministries at the national level and to serve as a gadfly to stimulate action along desired lines. For these purposes, the government created in 1963 the Delegation for Territorial Planning and Regional Action (DATAR). Since 1967, both DATAR and the General Directorate of Planning have been responsible to the Prime Minister through the authority of a special Minister. The functions of the Delegation touch all aspects of regional development. It maintains liaison with the regional agencies. It participates in the preparation of the regional portions of the national plan, in the preparation of the regional public investment budgets, and in evaluating possible modifications of the regional development policies proposed by the regional and national advisory councils. It presides over and provides secretariat services for the Inter-ministerial Committee for Territorial Planning and Regional Action which formulates basic government policy on geographic issues. It does the same for the Central Urban Planning Group, which brings together representatives of different ministries to deal with the problems of the principal metropolitan regions and which reports to the Inter-ministerial Committee for Territorial Planning and Regional Action. It is also responsible for the management of the Intervention Fund for Territorial Planning. This fund, amounting to 1½ per cent of the public investment budget (equivalent now to about $100 million annually), is essentially a "sweetener," to foot part of the costs of regional activities, programs, and studies and thus promote projects in high priority development areas which might otherwise be neglected.[12]

But all of this machinery would be of little value unless there were significant ideas to promote, coordinate, and implement. It was the Fourth (1962–1965) and Fifth (1966–1970) National Plans which spelled out these ideas. The emphasis in the Fourth Plan, in contrast to the efforts during the previous decade, was on the need to devise appropriate policies for all of the regions of France. The different regions did have their special problems and many of the measures taken to deal with them were bound to affect each other. For simplification, a broad distinction was made between the growing and the less prosperous regions. For the former, the Fourth Plan proposed complementary policies (*politique d'accompagnement*). These were to reinforce desirable aspects of growth, curb excesses or diseconomies, and correct maladjustments caused by declining industries which might otherwise be masked by the general expansion of the region's economy. For the less prosperous regions, the emphasis was on propulsive measures (*politique d'entrainement*). These were intended to reduce the disparities in employment opportunities, since these disparities were closely related to the disparities in levels of investment, cultural facilities, income, and migration patterns.

There still remained the question of how and where to allocate investment in the different regions. One answer—first suggested in a study by Hautreux and Rochefort—was to build up the metropolises outside of Paris.[13] The authors took as their basic starting point the view, expressed by the French geographer Pierre George, that in the past regions made cities but that today cities make regions. The main question raised was how to identify the principal metropolitan areas which could serve to develop their regions and to offset the dominance of Paris. This was a refreshing change from the essentially fruitless inquiries in the past concerning the optimum size of a city. It was taken for granted that a large city offered numerous advantages and that it would be useful to devise a pragmatic measure of the most significant French cities and the potential areas of their influence. Four basic criteria were used for their selection: the size of the cities; the services they had evolved for economic activities; the special services or functions performed by the cities; and their zones of influence. Twenty or more indicators were employed to identify and to weight these factors.

Eight leading cities passed this screen. Some, like Marseille and Lyons, exercised a dominant influence in their region, and on important secondary cities such as Grenoble in the case of Lyons, and Nice and Montpellier in the case of Marseille. In other regions, such as the Southwest and Brittany, there were dominant cities such as Toulouse, Bordeaux, and Nantes which exerted their influence imperfectly—presumably because of inadequate facilities, services, and communication linkages within and outside the region. In the North and the East, the larger number of cities limited the zone of influence or dominance of any single metropolis, though one or two cities at the frontier appeared destined to play a far more significant role with the emergence of the Common Market. Beyond the zone of influence of the regional metropolises, there was also a vast area where the facilities, the services, and the zones of influence of the communities within this area were considered stunted due to the dominance exerted by Paris. These included cities like Caen in Normandy, Rennes in Britanny, Limoges and Clermont-Ferrand in Auvergne, and Dijon in Burgundy. Finally, within the Paris region, there was also a ring of satellite cities

(Amiens, Tours, Bourges, and Orleans) of which one or all appeared destined to serve as limited subregional centers for the decentralization of activities from the capital city.

The leading metropolises—all located at the periphery of France—differed in size, in function, in physical form, in their range of influence, and in their needs. It was taken for granted that more extensive and detailed studies would be required to evaluate their problems and their potentials. What Hautreux and Rochefort suggested, however, was that their investigations provided the lead and the emphasis that should guide national strategy for regional and metropolitan development. It was in these metropolitan areas, they argued, that the major influence could be exerted on the regions. Therefore, it was in these metropolitan areas and their hinterlands that the major studies and programs should be developed which would build up a critical mass of economic activity and population, enrich their services and cultural development, and multiply their ties and channels of communication to their hinterlands and to the other regions of France.

The lucidity and simplicity with which the ideas were presented contributed to their appeal. The argument held out the possibility of developing a national strategy based on the needs of other regions while contributing to the solution of the problems facing Paris. It embraced all of France, yet suggested a limited number of regions and metropolitan areas where action could be initiated with effects that would penetrate the entire hinterland, including rural areas as well as secondary cities. The basic idea was relatively easy to explain and to popularize and it was sufficiently general and even vague enough to permit flexibility in adapting to criticism or practical requirements. All essential points were satisfied. One could point up its "scientific" origins and its national, regional, urban, and rural dimensions. It provided a lead for stress and sacrifice in resource allocation. It also satisfied diverse claims. The idea had appeal for the provinces; it served Paris; and it was presumably consistent with future urbanization trends; and, perhaps most important, it also squared with what intellectuals and decision-makers in Paris and the provinces thought the

285

provinces required. It is not surprising, therefore, that the idea of equilibrium metropolises was picked up by the administrative organisms and with minor adjustments adopted as the basic urban and regional strategy endorsed by the Fifth Plan.[14]

The new doctrine furnished some fresh rhetoric for national planning and some new leads for regional programming. The basic territorial problem, the Fifth Plan emphasized, was the appropriate distribution of activity between Paris and the other agglomerations. The doctrine identified the principal metropolitan areas in the different regions of France where growth was occurring or would be systematically encouraged. A number of examples were cited in the Plan to show how the development programs would be adapted to this end. Industrialization of the West was now to be encouraged principally in three major equilibrium metropolises, Nantes-Saint Nazaire, Bordeaux, and Toulouse. It was in these metropolises that physical development plans were to be prepared, universities, government research establishments, and technical training institutes established or extended, highway and air activity encouraged. It was in their hinterlands also (the Marais in the West and the rural areas of the Southwest) that major agricultural modernization and irrigation schemes were to be undertaken, that major regional parks and recreational zones were to be established. An equally impressive list of infrastructure improvements—roads, canals, port facilities, educational establishments, and urban renewal plans—were outlined for the East, in particular for the urban regions of Lille-Roubaix-Tourcoing, Strasbourg, Nancy-Metz, Lyons, and Marseille. Since Paris was the most important growth pole, its needs for roads, schools, housing, for general modernization within the city and region, and for the economic and social development of the existing cities and of new cities within its basin were also sketched in as part of the comprehensive effort to deal with urban development.[15]

A few tentative targets were set to indicate the reversal of trends that was sought. Despite the problems of housing in the Paris region, production was to be augmented in the other regions. By 1970, the proportion of housing financed in the Paris region was to be less than the 27 per cent which was set in the Fourth Plan for 1965. Similarly, the relative number of students in the University of Paris was to be reduced still further. There were 41 per cent in 1954 and 33 per cent in 1964. The aim was to cut this to 26.5 per cent by 1973. The same policy applied to state investment in research. Under the Fourth Plan, it was 58 per cent for the Paris region and Orleans, 22 per cent for the ten largest agglomerations, and 20 per cent for the other larger cities. Under the Fifth Plan, the aim was to change these proportions to 35-40 per cent, 50-55 per cent, and 10-15 per cent respectively.[16]

It took approximately a generation after the start of the French system of national planning to evolve a parallel system for territorial planning with explicit policies for urban and regional development. The main targets are the decentralization of Paris, industrialization of the West, the integration of regional and national planning, and the encouragement of modernization and new development mainly by focusing on growth poles in the major metropolitan areas in the East and West and around Paris. Although the progress of the programs left much to be desired, no alternative strategies were advocated. Opposition was negligible, even though opinion polls indicated that the French population generally—and the inhabitants of the equilibrium metropolises in particular—disliked big cities.[17]

Despite the apparent consensus, the new policies created serious strains. These challenge some planners, but they worry or discourage many others. The problems range from the need to create more effective machinery for regional government and local participation to the growing dismay because of the shortage of funds and staff, the conflicts among cities within and between metropolitan areas, and the inadequacy of existing tools for controlling the zooming prices of land or engaging in effective metropolitan planning and implementation.

The most persistent criticism throughout this period was the overweening centralization. There was no effort to set up autonomous regional authorities with legislatures and with independent revenue raising mechanisms. Critics, therefore, found it easy to show that local officials resented the limited attention paid to their views.[18]

287

There were also complaints about the effectiveness of the policy for relocation of industry. Decentralization of Paris lagged. Although the proportion of permits to build in Paris declined, the data do not cover smaller and average size firms; and therefore, there is some question as to how much growth has been arrested.[19] Moreover, most of the decentralization occurred in the ring of cities between 80 (50 miles) and 200 (125 miles) kilometers from Paris, where such decentralization was most likely without economic assistance. There is still another critical weakness. Key officials have observed that France (as of September 1, 1966) spent less than one-sixth the amount that Britain did for these purposes, and they expect no significant change in the situation until these premiums are raised.[20] The situation may even get worse because of the disadvantage which the Common Market imposes on the poorer regions and the metropolitan areas in the West because of their distance from the principal concentrations of population and income.

Perhaps the most serious practical problem is the allocation of infrastructure investments. The aim of the planners was to avoid the frittering away of resources by focusing development in growth poles. But the large number of growth poles and the wide range of possible investments in and among the growth poles (not to mention outside of them) has produced a sprinkling effect nonetheless. Thus, it has been virtually impossible to finance the new towns proposed for Paris, and the expansion of the cities in the basin of Paris and the center of France proposed in the Fourth and Fifth Plans, and also growth in the eight equilibrium metropolises—not to mention the secondary cities within these agglomerations—or other growing points such as Grenoble or Dunkerque which were likely to experience formidable growth because of special reasons.

The problems, however, can be exaggerated. Though the boat, as one official put it, is not yet moving in the right direction, it is definitely turning. Or, at any rate, so think the most optimistic of the planners. And this may be so. Despite the problems and inadequacies, available indices—tax returns, holdings of savings, energy consumption, as well as new industrial jobs—do indicate a relative improvement in the West in the standard of living and in the growth of industrial jobs compared to the rest of France, particularly for the period from 1960 to 1968.[21] Some of the credit for this may be due to the fact that the French planners have faced up more frankly to the need for linking economic and urban planning on the regional and national scale than any other Western country.

French regional planning policies have shifted from a focus on physical planning, decentralization, and *ad hoc* regional programs to a national system of economic, social, and regional planning with emphasis on development. New agencies or institutions have been devised to serve these ends. Leadership, too, has shifted from a sector agency—the Ministry of Construction and of Urbanism, the initial pioneer—to the national development agencies, thus reflecting the broadened conception and scope of urban and regional development. In addition, the French planners have experimented with an explicit national strategy for regional development and with new forms of regional organization and representation. Incentives, controls, infrastructure allocations, and other mechanisms are being adapted to serve these objectives. They have pioneered the regionalization of budget expenditure for their public investment programs. They have created and are creating new organizations to collect the basic data in the form required to carry out the necessary urban and regional studies. They are beginning to re-examine their antiquated system of professional education in urban and regional planning. They are, in short, encouraging change and mobility in many significant ways. What is more, the current ideas and machinery are still regarded as provisional hypotheses which are being tested and are likely to be improved over time. These are no small accomplishments.

## Notes

1. The author gratefully records his thanks to Peter Hall, Serge Antoine, Max Stern and Pierre Viot for their helpful suggestions.
2. The Home Counties include Bedfordshire, Buckinghamshire, Essex, Hertfordshire, Kent, Middlesex and Surrey. For data from 1801–1901, see *Royal Commission on the Distribution of Population,* Cmd. 6153 (London:

HMSO, 1940), Table p. 22. The ratios for 1961 are based on data in Ministry of Housing and Local Government, *The South East Study 1961–1981* (London: HMSO, 1964), Tables 3–6, pp. 120–123. This area extends up to about 40 miles from central London. See also A. F. Weber, *The Growth of Cities* (New York: The Macmillan Co., 1899), p. 46.

3. T. Wilson, *Policies for Regional Development*, University of Glasgow Social and Economic Studies, Occasional Papers No. 3 (Edinburgh and London: Oliver and Boyd, 1964), p. 62 and pp. 67–70.

4. W. F. Luttrell, *Factory Location and Industrial Movement* (London: National Institute of Economic and Social Research, 1962), Vol. 1, p. 300.

5. See, for example, South East Economic Planning Council, *A Strategy for the South East* (London: HMSO, 1967); and Standing Conference on London and South East Regional Planning, *The South East—A Framework for Regional Planning*, LRP 1180 Agenda Item 18, July 17, 1968.

6. The *Times* (London), January, 1966, p. 15. The *Times* reported this view expressed by an official of the North East Development Council.

7. F. Cairncross, "Grapeshot Regionalism," The *Times* (London), July 28, 1967, p. 21.

8. Statement made by H. F. R. Catherwood, Chief Industrial Advisor, Department of Economic Affairs at the Scottish Branch of the Confederation of British Industry, cited in The *Times*, (London), Feb. 1, 1966, p. 15. See also *Investment Incentives*, Cmd. 2874 (London: HMSO, 1966); see also The *Times* (London), July 28, 1965, p. 10; Department of Economic Affairs, *The Development Areas—A Proposal for a Regional Employment Premium* (London: HMSO, 1967).

9. The *Times* (London), Feb. 1, 1966, p. 11. See also *The Development Areas, op. cit.*, p. 10.

10. J. R. James, "Regional Planning in Britain," in S. B. Warner, Jr., ed., *Planning For a Nation of Cities* (Cambridge: The M. I. T. Press, 1966), p. 206–207. The *South-East Study* (referred to an area including East Anglia. A more recent report from the South-East Council, *Strategy for the South-East*, forecasts an extra 2.62 million people in the same period for a smaller area which excludes East Anglia.

11. M. Astorg, "Sur Les Principes et Les Structures De La Planification Régionale" (Exposé presente par la Délégation française, les 5 et 6 Octobre 1965 et résumé des discussions auxquelles ils ont donné lieu, Groupe De Travail du Comité de l'industrie) (Paris, Organization of Economic Cooperation and Development, December 30, 1965), pp. 39–59; J. Lajugie, "Aménagement du territoire et développement économique régional en France (1945–1964)," Développement Économique Régional et Aménagement du Territoire, *Revue d'Economie Politique* (January-February, 1964), pp. 278–336; see also P. Viot, "Regional Aspects of French Planning," unpublished paper for Regional Planning Conference, sponsored by the National Institute for Physical Planning and Construction Research, May 1965, p. 12; Ecole Nationale d'Administration, L'Adaptation de L'Organisation Administrative Française, aux Prob-

lèmes D'Aménagement des Grandes agglomérations urbaines, Séminaire dirigé par P. Viot (Paris: Mars, 1966), pp. 19–46; and J. and A. M. Hackett, *Economic Planning in France* (London: G. Allen and Unwin, Ltd., 1965), Chaps. IV and XIII.

12. Projets de Loi De Finances pour 1966. Pour 1967, Pour 1968. Execution du Plan, Aménagement du Territoire, Regionalisation du Budget. Imprimerie Nationale.

13. Jean Hautreux and Michel Rochefort, *La Function Régionale Dans L'Armature Urbaine Française,* Commission Nationale de L'Aménagement du Territoire (Group V), Commisariat Général du Plan et Direction de L'Aménagement Foncier et de L'Urbanisme, Ministère de la Construction, Paris, April, 1964.

14. For a presentation of the main elements of the official policy, see $V^e$ (Fifth) *Plan De Développement Economique et Social* (1966–1970), Tome I, pp. 131–134; Commission Nationale de L'Aménagement du Territoire, Premier Rapport, Commissaire Général du Plan D'Equipement et De La Productivité, Sept., 1964, pp. 81–91. See also Métropoles d'Equilibre, in *Urbanisme,* 89, Paris, 1965; and O. Guichard, *Aménager La France,* Paris Editrous Gonthier, 1965; and J. F. Gravier, *L'Aménagement du Territoire et L'Avenir Des Region Françaises,* Paris: Flammarion, 1964.

15. Fifth Plan, 1966–1970, Vol. I, *op. cit.,* pp. 131–134.

16. *Ibid.,* pp. 132–133.

17. See, for example, A. Girard and H. Bastide, "Les problèmes demographiques devant l'opinion," *Population* (April-May, 1960), pp. 246–287.

18. P. Gremion, *La Mise En Place des Institutions Régionales,* Centre de Recherche de Sociologie des Organisations, Paris, 1965; and J. P. Gremion and J. P. Worms, *Les Institutions et la Société Locale,* Groupe de Sociologie des Organisations, Centre National de la Recherche Scientifique, Paris, 1968.

19. Ministère de la Construction, *Bulletin Statistique Mensuel,* (July-August, 1965), pp. 98–101, and Projet de Loi (1966), Vol. II Tome, I, *op. cit.*

20. P. Viot, *op. cit.,* p. 20.

21. Projet de Lois de financer 1966–1967–1968. (The government publishes each year a report on the execution of the National Plan, Regionalization of the State Budget of Regional Actions.)

# 23 URBAN RENEWAL AND AMERICAN CITIES

## Scott A. Greer

When we look at our current efforts to renew American cities, it is important to remember how we made these cities in the first place. We had a vast, unexploited continent, rich in the resources that sustain human life; it was a fortunate time to enter history, when economic, political, and transportation freedoms were expanding; most important of all, perhaps, there was a growing faith that men could expect untold wealth from the maximum freedom to explore economic possibilities with no bows to the idols of the tribe or the monarch—the only limits being those required to preserve individual freedom and decency.

Under these circumstances American cities grew at the crossroads of transportation. On the Atlantic seaboard they were entrepôts, gathering spots for wealth traded to Europe for luxuries and the accoutrements of civilization. Inland they arose where the waterways crossed the great transcontinental trails (which we owed to the American Indians) or converged upon the great inland seas, the Great Lakes. The center of the city was always the place where the cargoes met and the goods were transshipped. The city developed outwards from this center, in a roughly concentric pattern, based chiefly upon the cost of moving men, materials, and messages. But the distance which this growth could spread was severely limited by the cost of movement; thus the cities tended to be limited horizontally. They grew upwards, especially at the center.

Because central location was highly valued, since it made movement cheaper, land at the center had a great and persisting value. This was reflected in the market price for central land, since it meant advantages in the markets for labor, capital, and resources. The center could be expected to maintain its value indefinitely. There would be an automatic and continuous rejuvenation of the center; its structures would be forever rebuilt, for it had an ineluctable advantage over the peripheries. It was closest to most of the things that mattered. (Such a situation exists today in many societies, where transportation is roughly similar to that in the United States of the early nineteenth century.)

This was the situation at the time of the American Revolution. In the next century things changed dramatically, with the appearance of a companion revolution in energy resources—the Industrial Revolution. With the application of steam power to human work the leverage of the human hand increased enormously; production, fabrication, and transportation could all be multiplied by a very large factor. But the increased power in transportation allowed by the steam locomotive did not alter a basic fact: while the movement of massive, heavy cargoes between cities grew cheaper, movement within cities continued to be costly and difficult, for the heavy trains could not be economically used within cities. Thus cities grew in population and in wealth, but tended to remain highly centralized (though now where rail and water converged), densely built, and highly structured.

## A BRIEF HISTORY OF URBAN THEORY

Social science theory of cities tends to be very biased by the particular period in which the scientists work. This is not surprising; recorded human history (like cities) is only some seven thousand years old, and there are not many repeating cases. (Our fascination with primitive societies is in part due to the fact that they are similar and numerous because primitive; they are not historical.) Furthermore, a great step function like the exploitation of coal through the use of the steam engine, and the following effects on production and transportation, was quite unprecedented. Thus men studying urban society in the latter nineteenth and early twentieth centuries tended to take what they saw for granted: they concretized a given state of things. They built the crude beginnings of urban social science. How did they do it?

They began with observations of land use; from real estate research (a major concern of Americans, who are all speculators in land), and from the mapping of populations and activities over space, they developed a "folk-map" of what the city is. It is, they said, a set of concentric circles, with transportation crossroads at the center, heavy industry nearby, the market-place and homes of the workers close at hand, and, as we move outwards, increasingly wealthy residential neighborhoods. For distance from the center means increased transportation cost and, therefore, increased ability to pay—wealth.

Thus the image of the city created by early twentieth-century social science was comparable to a cross-section of a plant where growth occurred at the center, with new commerce and industry inexorably moving outwards (though also upwards because of the pressure of space). This growth intruded on older, residential neighborhoods, slowly spoiling them for residential purposes but ripening them for new uses. They were future sites for market and factory. The increase in economic activity, resulting in population growth, meant continuous new building on the peripheries—but only within the limits of transportation costs. The metropolis was held together by the great linchpin of the market and the factories near the center; here the jobs were, the wealth and the excitement, the things that held a spatial dispersion together as a single community. This was the city of the railway age, a city which lasted well into the 1920's.


## THE GREAT DEPRESSION AND THE PUBLIC RESPONSE

The great economic Depression of the 1930's meant a slowing down of all economic growth in America. Construction faltered and stopped in many central business districts; housing was sold for taxes due, while families doubled up; the construction industry suffered acute depression. In general, the public spaces and structures of the central city stagnated, the business buildings grew old, and the results were unseemly. The residential areas around the center also grew older, for there was little capital to invest in maintenance.

Early in the Depression, President Roosevelt's New Deal came into power. It was an administration basically committed to amelioration and reform within the framework of democratic capitalism—one which allowed the maximum of freedom to private entrepreneurs within the limits, but also with the aid, of the public power and purse. New Deal thinkers and leaders saw a sick construction industry, with inadequate capital to build for whatever market existed, and a deteriorating inner city, with slums growing faster than new construction.

The intellectual response to these circumstances was fairly complex. It was assumed, first, that an important objective was, in Charles Haar's words, to get "builders to build and lenders to lend"—that is, to stimulate credit in housing. The result was an act to provide federal insurance for home-owners' mortgages. This act, however, left discretion as to lending in the hands of the traditional financiers of house construction, the banks and building and loan associations. Second, it was assumed that the declining residential neighborhoods in the central cities, the "slums," were the result of simple inability to pay the price of good housing. Thus the program of slum clearance included the replacement of every destroyed substandard house with a new, publicly subsidized housing unit.

Finally, it was assumed that the decline in the vitality of the central commercial and industrial district was due to obsolete landownership patterns. Thus one could encourage the rebuilding of central business districts through facilitating the merger of small plots of land, clearing them of old and largely obsolete structures, and allowing for new building at the center.

The result was a series of legislative acts. All were based upon the social science theories described earlier. The housing act of 1937 provided for federal subsidy, through mortgage guarantee, of new building; this would presumably take the shape of new building within the historical pattern of the city. The central core, the business district, would continue its growth outwards, once the "artificial" handicap of obsolete land-tenure patterns was cleared away. The slums would disappear with the destruction of poor housing for poor people, and the substitution of good (though subsidized) housing for poor people. (It was, incidentally, assumed that all of these measures were temporary—emergency measures created by the Depression.)

The remedies of the Depression legislation were never thoroughly tested, for World War II intervened. Although the war greatly increased economic productivity, it virtually halted construction of domestic structures. Thus, by 1945, American cities had had their housing demand immensely increased over a fifteen-year period, with only a puny amount of new housing built to meet that demand.

## THE POSTWAR PERIOD

After World War II, the United States entered into a period of unprecedented (and unpredicted) economic growth. The enormous productive apparatus created with public subsidies for wartime tasks turned smoothly to producing for the domestic consumer market, and the nation was rapidly transformed into something it had never been before.

The introduction of the truck and the automobile had been a force in the restructuring of cities during the 1920's and 1930's. This had been masked, however, by the general Depression and the war. Now automobiles and trucks were built in enormous quantities and sold to a nation that could afford them, which had the effect of greatly expanding the distance over which the urban population could spread while remaining effectively integrated. Thus the city that emerged after World War II differed drastically from that postulated by the older urban social science.

It is a city of rapid and startling decentralization of structures. The residential neighborhoods spread outwards into the fields and orchards of the countryside, in some cases (as in Los Angeles County and the Niagara Peninsula of Ontario) pre-empting valuable agricultural land. They are made possible by the great networks of freeways and the automobiles that use them; they are accompanied by the development of enormous shopping centers. And, as the trucks pre-empt a large proportion of intercity transport of goods, the industrial complexes move outwards to the peripheries. Thus today many industries deal from the circumferential highways and industrial parks of one metropolis to their twins in other areas; the older center, where rail and waterway meet, is of little import.

Such a city is best exemplified by Los Angeles, one of the new cities of the west. Built after the automobile became current, assuming the great highway gridwork, the truck, and decentralized commerce and industry, it is less an urban form than an urban texture. With a density lower than that of Java, it is yet one of the most affluent, energetic, and large-scale cities on earth. In a sense, it has no center; instead, it has a dozen centers scattered over 3,500 square miles. So different is it from the city of the railroad age that many people, who have in mind a New York, Chicago, Paris, or Rome, assert it to be no city at all. Yet it is probably the American city of the future.

## URBAN POLICY SINCE WORLD WAR II

The policy of the United States government since World War II has been largely devoted to restoring the city of the railroad age. Still operating with the primitive urban social science described above, policy makers have continued to attempt to restore the vitality of the central city and to "wipe out the slums" through destruction of ugly neighborhoods. This set of policies is know as urban renewal. It is an intricate program which has snowballed, from an original concern (in the Depression) to get decent housing built for the population, to a concern with revitalizing the central city, to a concern with planning the growth and future of the metropolis as a whole. There have been fatal flaws in each aspect of the program;

they stem from the basic nature of the urban renewal laws, and, finally, from the nature of political culture in the United States.

The effort to improve housing in the city has been based upon the notion that the destruction of slum housing would result in an upgrading of housing for the poor. In brief, laws would force landlords to keep a minimum level of decency in housing, under penalty of having it condemned and destroyed, while the building of new housing would allow older (but better) housing to "filter down" to the poor. But the newer housing does not filter down, for there is too much difference in price, as William Grigsby has shown in *Housing Markets and Public Policy* (1963). Meanwhile, destruction of cheaper housing has simply made a tight housing market for the poor tighter still—allowing slumlords to charge more for less. As this becomes evident, opposition to slum clearance begins to develop among those most injured—the segregated and the poor as well as their liberal allies.

The destruction of cheap slum housing can be justified only when it is replaced by decent housing which can be afforded by those displaced. This means, of course, some form of subsidized housing: poor people living on minimum incomes cannot afford what is considered standard housing by the middle-class public servants who make and enforce the law. The goal of revitalizing the central business district has gotten the urban renewal program considerable support. There is an old tradition of beautifying the city in the United States; it is closely integrated with the goal of continuous expansion of the city, or "boosterism." And it serves interests, financial and otherwise. It focuses on the core of the city, where banks, retail stores, newspapers, transportation media, and headquarters offices are located. The core is of structures from one part of the area to the other. But the bribe is really inadequate; the price of land is not that important a determinant for the kinds of buildings wanted in the core city. Central city revitalization programs usually either reclaimed for expensive use land that would have been reused anyway or produced a drug on the market. For if there is no great demand for center city space, creation of new structures merely siphons off tenants from older ones and, in the process, lowers their occupancy rate and quality of maintenance—in short, spreads blight. If demand for central city space is limited, building new structures merely pours water from one pitcher to the other.

The notion of planning the growth and future of the metropolis as a whole developed as it became clear that the metropolis was, in ways important to urban renewal, a unit. It was one space market—new housing on the peripheries competed with older housing in the center; suburban shopping centers competed with the central business district; suburban industrial parks took plants away from the congested, older industrial areas. Thus the notion of reshaping the central city led to the notion of planning for the city as a whole over time.

But the goal of planning, the comprehensive renewal plan, runs into several constraints built into the very structure of American society. The right to free incorporation of municipalities means that most metropolitan areas are divided into a large central city, including perhaps half the interdependent populations, and dozens or hundreds of independent suburban municipalities. The programs for renewal are applicable only to a given, responsible municipal unit: thus they cannot embrace the total area to be considered, planned, and legislated for. So feeble is the coordination among metropolitan governmental jurisdictions that there is rarely even information on the relevant facts, much less ability to plan for the metropolitan area as a whole.

An equally important limitation on comprehensive planning is, simply, the political weakness of city planning in the United States. Having no ability to build new structures (beyond advisement), and no control over the structures created by private enterprise (beyond temporary obstruction), city planning is a weak and ineffectual control mechanism in American cities.

Limited thus in both areal jurisdiction and real power to control, the effort at comprehensive urban planning amounts to little more than utopian rhetoric. Meanwhile, at the peripheries, growth goes on at a rapid pace, new suburbs, shopping centers, and industrial parks spring up, and the older center is left behind to slowly develop its own form of working-class society—partly black ghetto, partly the abandoned poor whites, the old, the crippled, and disorganized and partly, of course, the productive and increasingly prosperous industrial working class and lower-middle class. The latter live, for the most part, far from the center in the "outer wards"—those most like sub-urbia.

## WHAT IS TO BE DONE?

The American metropolis is caught in a tug of war. On the one hand there is the demand to exercise the right of free enterprise—to make a fast buck if possible. This results in the uncontrolled exploitation of urban land, used or new, controlled only by someone's balance sheets and the tax dodges possible to improve those balances. On the other hand, several centuries of such exploitation have produced a kind of metropolis which is, in the eyes of many, costly, inefficient, and ugly.

Some would like to recreate the centralized city of the railroad age, a clear-cut form with center and centrality. Others would like to realize all the values possible in the decentralized urban fabric, the city of a hundred centers, the city of dispersion. Neither can succeed while the public power is so limited and the market in capital and land is free to determine new growth and, through lack of maintenance, the decline of old growth. However, as long as such freedom is allowed to speculators and entrepreneurs, one would have to bet on the dominance of the decentralized city and the decline of the old centers.

But we as Americans are caught in another tug of war, that between the importance of the national purpose and the autonomy of the local community. It is not unlike the first, for it also reflects the desire to have free enterprise in government (any small population settled near one another have a right to self-rule) and the overriding national purpose. In general, the local autonomists are in control; suburban municipalities burgeon while national policy assumes integrated metropolitan communities.

But until national purpose can become effective in the cities, we cannot expect any given city to assume control of its destiny. The cities are in competition with one another—for capi tal, labor, and markets. With the free movement of these factors we have what Norton Long in *Polity* has called "civic mercantilism"—competition for scarce resources, which results in price cutting. The cities cannot demand enough of their citizens because their citizens are living in what Morris Janowitz called "communities of limited liability." Individuals and corporations alike can move across boundary lines to places where public law is less demanding, and more important, new growth can move outside the existing corporate boundaries.

In truth, to have a real policy for a given city we must have a national policy for all cities. We must include the entire playing field in our jurisdiction, considering all uncommitted land a national resource, and a precious one at that. Looking at the projections of scores of millions of new urban Americans by 1999, we must take care that the growth of structures which will house them and help them earn a living does not strangle their personal lives, create chaos in their social order, and foreclose their freedom to choose. We cannot do this through the free and easy exploitation which made shacktowns out of the prairies and jerry-built slums along the estuaries of our great polluted rivers.

# 24 NEW TOWNS

Edward P. Eichler
Bernard Norwitch

In every kind of journal, from provincial weeklies, *Life,* and *Realities* (*Réalités*) to publications of the United Nations, the idea of "new towns" has been discussed and acclaimed as a solution to the world's urban dilemma. Why is this so? What were the origins of the idea? What are the urban ills new towns are to cure? Is the concept applicable both to industrialized and developing societies? Who does and who should initiate new towns? It is our aim to answer these questions, but first a brief historical sketch is in order.

## A WORLD OF NEW CITIES

All urban settlements were once "new." They emerged from folk societies some 5,500 years ago, when selective cultivation of grain and the practice of animal husbandry resulted in an agricultural surplus. A food surplus permits both labor specialization and a class structure that can, for instance, develop and maintain an irrigation system. As far as we know, the world's first cities were formed around 3500 B.C. in the Tigres and Euphrates valleys and shortly thereafter in the valley of the Nile. In these, as in most early cities, temporal and spiritual leadership was vested in one man or group. The king or the priest received part of the food surplus. In return he offered protection on earth and peace after death. With the surplus he paid the cost of protective walls, supported an army, and built houses of worship, all of which constituted the infrastructure of his city.

As technology, both in production and transportation, improved, specialization could increase and trade within and between cities expanded. The development of money and of literacy furthered this expansion. The earliest cities were founded where the soil was especially fertile. But as the possibility of intercity trade arose and as rulers like those of Rome sought to expand their empires, proximity to waterways and ease of defense became equally important criteria. In addition, cities were founded at or between the sources of valuable raw materials like silk and rubber.

Thus, before the Industrial Revolution, the location of cities was dictated by their proximity to fertile soil and waterways, the distance between such places (as measured by the travel time for animal-drawn vehicles), by the ease of warding off enemies, and by the existence of natural resources which could be tapped and transported with limited machinery and power. These urbanizing forces and criteria, important though they were, affected only a small portion of the world's population. According to demographer Kingsley Davis, by 1800 only 2.2 per cent of Europe's population lived in cities over 100,000. Because industrialization began earlier in England, the percentage there in 1800 was 10. In forty years it doubled and in sixty years doubled again.

The development of steam and then electric power and the creation of new kinds of legal and social arrangements permitted a degree of specialization of labor formerly thought impossible. Rail and later motor transportation opened vast areas for exploitation despite their distance from the sea. The demand for raw materials like coal and iron ore and for hydroelectric power created such great urban, industrial areas as Germany's Ruhr, the United States' Midwest, and England's Midlands.

Not only did the Industrial Revolution vastly increase urban specialization, it created an agricultural revolution. The nineteenth and the first half of the twentieth centuries saw Western nations, which even 5,000 years after the formation of the first cities had had over 95 per cent of their people living on farms, become predominantly urban.

Specific topographic conditions, which once mattered so much for defense, had been rendered irrelevant by modern weaponry. But public policy which once had created cities as centers of religion now had a new spiritual drive—nation-building. In the United States the great nineteenth-century expansion from the East to the hinterlands was driven not only by the existence of raw materials and fertile soil but also by the spirit of "Manifest Destiny." Although the stated reasons may have been economic, nations in reality fought over and settled territory as a matter of national pride.

## THE RESPONSE

Storage, defense, political and religious ritual, trade—these had been the reasons for the creation and growth of new cities and the criteria for their location before the nineteenth century. With industrialization came new reasons and criteria—electric power, proximity to other fuels and raw materials basic to mass production, and nation building. With all of the addition to wealth and the opportunity it offered for rural inhabitants to move to the city, industrialization, as people relentlessly sought that opportunity, dramatized the disadvantages of large cities.

By the end of the nineteenth century a growing number of observers were becoming shocked at the congestion, pollution, and apparent social breakdown caused by the flight from farm to city. Even before the end of the century Dickens portrayed these conditions with outrage. A few men, mostly in England where industrialization and city growth were most advanced, began not only to declaim the physical, social, and moral conditions of big cities, but to suggest solutions. One of these solutions was the construction of new towns. If for 5,000 years monarchs and businessmen had started cities for money and glory, why shouldn't they now build towns to improve the quality of the environment? It was this idea of "quality" and the nineteenth-century notion that all problems could be solved by rationalization which were the basis for, and the novel aspect of, the concept of new towns. Although an accepted necessity, the city was seen as evil, at least in part, and some kind of medicine was needed to counteract the malignancy.

Modern American thought about the defects of contemporary urban life and the social organizations in which these defects could be righted begins with an Englishman, Ebenezer Howard. Writing in the last decade of the nineteenth century, Howard based his view of contemporary urban life on an analysis of the waste and disorganization which the Industrial Revolution had brought to Europe's major cities. Howard saw urban centers growing larger and larger and felt that this growth would intensify all the problems of the city and make life there less and less humane. In his book, *Tomorrow: A Peaceful Path to Real Reform,* first published in 1898, Howard proposed that the English government establish a series of small, self-sufficient towns under public control. The population of each was to approximate 30,000. By owning the land, the town could profit from the appreciation in land value and thus finance local services. The towns would be connected by transportation systems to the country's major urban center, London, and would be designed to catch London's "overspill." Each town would be protected from encroachment (and prevented from expanding) by a permanent greenbelt circumscribing its borders.

Howard did not conceive of the new towns he proposed as elements that in themselves would right the defects of city life. Rather he saw a symbiotic relationship between city and suburb:

"There are in reality not only, as is so constantly assumed, two alternatives—town life and country life—but a third alternative in which all the advantages of the most energetic and active town life, with all the beauty and delight of the country, may be secured in perfect combination; and the certainty of being able to live this life will be the magnet which will produce the effect for which we are all striving—the spontaneous movement of the people from our crowded cities to the bosom of our kindly mother earth, at once the source of life, of happiness, of wealth, and of power."[1]

Howard's concept of the garden city found many adherents in the United States, one of the earliest being Patrick Geddes. In *Cities in Evolution,* written a decade after Howard's work, Geddes took the garden city concept and proposed such towns as part of regional plans. The first realization of Howard's concept in England came in 1904 with the developments of Letchworth (designed by Barry Parker and Raymond Unwin) and Welwyn (Unwin). For half a century, city planners and intellectuals in increasing numbers advocated new towns, but international strife and economic instability occupied the minds of Western European governments. In the United States, three new towns were started in the 1930's, but this was one of a number of scattered acts to combat the Depression and not essentially a program to improve the quality of the urban environment.

The end of World War II set the stage for a wave of town building. The intellectual arguments had long since been mounted. New towns would offer a safer, cleaner, saner, more efficient, and more democratic environment. A smaller town, it was thought, would allow people to identify with it and thereby commit themselves to its government. If the war had been fought to preserve democracy, was this not the time to make self-government operational? The argument was especially compelling because the war's destruction required massive rebuilding in any case.

Led by England, whose Parliament passed the New Towns Act of 1947, European nations like Sweden, Holland, and even the Soviet Union built new towns after the war. In England and the Soviet Union the towns were planned as related to, but still separate from, existing cities and were to be limited to 60,000 to 100,000 people. They were to have enough industry so that few, if any, residents would be required to commute to jobs outside the town.

In almost every instance the towns were initiated and financed by the government. The land and facilities were owned by local government or a public corporation chartered for this purpose. Their growth was aided by governmental policies regulating industrial location. In some cases the land had to be purchased from private owners (England), but in the Soviet Union and Sweden the government already owned the land. The underlying thesis of this movement was that a better relationship of urban functions such as housing, commerce, industry, education, recreation, transportation, and culture could be organized by starting fresh. In addition, people and industry were to be decanted from major metropolitan areas like London to relieve them from the burden of excessive growth. In developing nations like those in Eastern Europe and in poorer countries like Venezuela, new towns were also started to aid in the exploitation of resources such as iron ore and oil.

We do not have the time to catalogue completely the benefits which new towns in developed nations were to confer on their occupants, the nearby metropolis, and the nation as a whole. Certainly one was the chance to relate urban growth and transportation lines more efficiently. Another was to preserve land for recreation and open space between towns and the city. A third was to have society as a whole benefit from land values increased by urbanization.

Underpinning these "rational" arguments has been an aversion to very large cities—to the metropolitan and megalopolitan character of modern urban growth. To the most ardent advocates, the word "towns" has been especially significant as it conveys the conviction that urban areas of moderate size are virtuous—economically, politically, and morally. The English and the Russians have failed to limit the size of their major cities, despite myriad plans and controls, and have raised the maximum population of some new towns from 50,000 or 60,000 to upwards of 150,000. To many planners and philosophers, this has represented not intelligent adaptation but retreat in the face of the enemy's onslaught.

Since the middle of the 1950's, academic, professional, and popular journals have increasingly portrayed the horrors of metropolitan expansion in the United States. Lewis Mumford, Clarence Stein, and others had initiated such criticism thirty years earlier, drawing upon the ideas of Howard. Yet, here in the world's most urbanized (over 70 per cent of the population live in urban areas) and affluent nation no governmental pro-

gram to initiate new towns has been enacted. The reasons may
be found in the political and philosophic origins of the United
States.

The United States is the only major nation of the world
which did not pass through a long period of despotic and
feudal rule. Its founders came to escape nations where the
king was still powerful. They rebelled against the crown. All
of its early heroes like Washington, Jefferson, and Franklin
were investors and speculators in land. Its Constitution was
carefully drawn to balance power, to protect individual liberty,
and to limit the national government's power over property.
In major part the Union was formed to establish a favorable
climate for industry and trade. Even at its inception the business of the United States was business.

Throughout the nineteenth century the federal government
disposed of its vast land holdings to encourage settlement of
new territory. Railroads and homesteaders were two of the
major recipients. Growth, expansion, exploitation of resources
—these were the driving forces of nineteenth-century America.
European nations, whose cities were full of grand parks, plazas,
and shrines built by kings and clerics, respected the land. Even
their businessmen acknowledged the role of the state as the
dominant force in controlling development. The Swedish Crown
gave its land to the city of Stockholm so that the city government could guide its growth. The United States government
gave its land to war veterans and sold it to speculators. Each
nation is a democracy, but history provided for each a very
different basis for making decisions about land use.

One aspect of the intellectual, moral, and political climate
of the United States, however, has made it susceptible to the
idea of new towns. A great many of us, especially our intellectuals, have distrusted large cities. Thomas Jefferson was only
the first of a long list of prominent American critics of the
city. Emerson, Thoreau, Melville, Poe, and John Dewey (in his
later life) all found the American city distasteful. In one sentence Dewey summarized the prevailing fear of many Americans, especially those interested in the quality of urban life:
"Unless local communal life can be restored, the public cannot

adequately resolve its most urgent problem: to find and identify itself."

Thus, if new towns are an antimegalopolis concept whose implementation would allow most of us to avoid the dirt, noise, confusion, and degradation of vast cities, we in America, at least nostalgically, are all for it. But if, in order to have new towns, we must grant to the state, especially the federal government, the power to control, own, and develop land, and the regulatory devices to channel growth to that land, we resist. These conflicting feelings seem irreconcilable, but Americans are not easily put off. If private ownership of land and freedom of movement are sacrosanct and if new towns are needed, there is only one answer: private industry, with little or no aid or control from government, must build them.

## THE AMERICAN EXPERIENCE

After World War II, American suburbs burgeoned. First came houses, then shopping and other service facilities, then industry. The development of these suburbs was done more and more by specialists. Merchant builders purchased 50 to 100 acres of land, installed site improvements, secured interim and permanent financing, set up crude but efficient assembly lines for mass production, and merchandized housing to the consumer. Others concentrated on the creation of neighborhood and regional shopping centers and industrial parks.

These entrepreneurs had to secure the approval of local governments for the size of the lot and the type of land use. Much of the profit in land development came from changes in zoning to use for smaller single-family lots, apartments, industry, or commerce. As competition increased and the postwar housing shortage was relieved, developers began to install and merchandise community facilities like swimming pools, meeting rooms, and tennis courts. All of this required planning and considerable capital, but few of these builders thought of themselves as engaging in a public-spirited venture. Their aim was to offer a product, which could be sold or rented as rapidly as possible, and to secure financial return commensurate with their investment of time and money.

In 1958 the first contemporary American new town was announced. It was El Dorado Hills, 10,000 acres thirty miles north of Sacramento, California. It was quickly followed by announcements of the purchase of 7,000 acres by Robert Simon for Reston, Virginia, and 15,000 acres by James Rouse for Columbia, Maryland. At the same time several families who had owned thousands of acres of land for fifty to a hundred years declared their intention to build new towns. The most dramatic of these is the Irvine Ranch, 100,000 acres at the southern end of metropolitan Los Angeles.

The American new towns are unlike those in England in several ways. First, they are initiated by businessmen, not government. Second, they are at the fringe of the metropolis, not beyond it, and there is no governmental provision for a separating greenbelt. Third, since the government does not direct industry to locate in them, they must depend primarily on commuters for initial customers. Fourth, there are no special governmental subsidies for the purchase or rental of housing.

Thus, American new towns are a different way of developing the metropolitan fringe, not an attempt to change the megalopolitan character of the nation's urban growth. Instead of ownership and development of modest sized parcels of land by specialists, American new town builders propose to plan and develop land for housing, commerce, industry, and recreation under one aegis. The result is to be a more sensitive and orderly use of vast tracts of land as compared to what we have seen in the twenty-five years since World War II.

## DISCRIMINATION

Unfortunately, most of the commentary on new towns has failed to discriminate between different cultures and different kinds of "urban problems." To the degree to which American suburban development has been disorderly and insensitive to amenities such as trees, stream valleys, rock outcroppings, and so forth (and we think it has), new towns, American style, offer a welcome promise. But to the problems of American urban life such as air and water pollution, traffic congestion, the deprivation of those whom Michael Harrington has characterized as the "other Americans," and relations between whites and blacks, new towns are irrelevant. To date they have been far less than successful financially, but this is more a subject for analysis at schools of business administration than a matter of national or international consequence.

Neither the American nor the Western European experiences with new towns have any necessary relevance to the problems of underdeveloped or developing countries. For them governmental promotion of the development of their hinterland may be desirable to exploit untapped natural resources and to infuse national spirit. To date Brasilia has been a failure as a national capital. But in the long run Kubitschek's dream of opening up Brazil's interior, with its vast natural wealth, may be realized and Brasilia may be seen as a decisive element in that realization. The great dilemma of the underdeveloped world is that it must accomplish in decades what a few nations did in centuries. The development of a wholly new system of communications has made tribal societies suddenly aware of the material benefits of industrialization. In the West, people migrated to cities as technology provided a surplus of food and opportunity for labor in cities. But a nation like India has not created the cultural or technological basis for such a migration. Yet, its people flock to its cities, where life may be even worse than it was on the farm.

If new towns are an appropriate device for such nations, it will only be as part of an over-all effort to develop the agricultural, technological, and political skills and institutions which are necessary to increase productivity. There is nothing in the Western experience with new towns which suggests that we have much to contribute to this enormous enterprise.

If we wish seriously to address ourselves to the major physical and social problems of American cities, we might start by pledging not to seek panaceas like new towns. Precision of thought and the will to do what is needed are essential. The poor are, in fact, poor. The unemployed are, in fact, out of work. Negroes are, in fact, discriminated against and poor. The air is getting dirtier. Certain parts of certain cities are overcrowded with people and cars. To those in underdeveloped countries we should show humility and sympathy. We do not have simple answers to their problems. We may be peculiarly ill-equipped to advise or aid them. Proposing British style new towns as a great cure-all for the urban problems of India or Nigeria is right out of Graham Greene's *The Quiet American.* Vietnam may have taught us how little we know.

## Notes

1. Ebenezer Howard, *Garden Cities of Tomorrow* (London: Faber and Faber, 1914), pp. 45–46.

# 25 THE CITY IN CHASSIS

## Daniel P. Moynihan

There is to be encountered in one of the Disraeli novels a gentleman described as a person "distinguished for ignorance" as he had but one idea and that was wrong. It is by now clear that future generations will perforce reach something of the same judgment about contemporary Americans in relation to their cities, for what we do and what we say reflect such opposite poles of judgment that we shall inevitably be seen to have misjudged most extraordinarily either in what we are saying about cities, or in what we are doing about them. We are, of course, doing very little, or rather, doing just about what we have been doing for the past half century or so, which can reflect a very great deal of activity, but no very considerable change. Simultaneously, and far more conspicuously, we are talking of crisis. The word is everywhere: on every tongue; in every pronouncement. The president has now taken to sending an annual message to Congress on urban subjects. In 1968 it was bluntly titled, *The Crisis of the Cities*. And indeed, not many weeks later, on Friday, April 5, at 4:02 p.m., to be exact, he was issuing a confirming proclamation of sorts:

> Whereas I have been informed that conditions of domestic violence and disorder exist in the District of Columbia and threaten the Washington metropolitan area, endangering life and property and obstructing the execution of the laws, and that local police forces are unable to bring about the prompt cessation of such acts of violence and restoration of law and order . . .

The excitement is nothing if not infectious. In its most recent publication, "Crisis: The Condition of the American City," Urban America, Inc. and the League of Women Voters noted that during 1967 even the Secretary of Agriculture devoted most

313

of his speeches to urban problems. In mid-1968, the President of the University of California issued a major statement entitled, "What We Must Do: The University and the Urban Crisis." The Bishops of the United States Catholic Conference came forth with their own program, entitled "The Church's Response to the Urban Crisis." At its 1968 convention, the Republican Party, not theretofore known for an obsession with the subject, adopted a platform plank entitled "Crisis of the Cities," while in an issue featuring a stunning black coed on the cover, Glamour magazine, ever alert to changing fashion, asks editorially the question many have posed themselves in private—"The Urban Crisis: What Can One Girl Do?"

Academics who have been involved with this subject might be expected to take some satisfaction that the alarums and jeremiads of the past decades seem at last to have been heard by the populace, and yet even those of us most seized with what Norman Mailer has termed the "middle class lust for apocalypse" are likely to have some reservations about the current enthusiasm for the subject of urban ills. It is not just a matter of the continued disparity between what we say and what we do: it is also, I suspect, a matter of *what* we are saying, and the manner of our saying it. A certain bathos comes through. One thinks of Captain Boyle and Joxer in that far-off Dublin tenement of Sean O'Casey's *Juno and the Paycock;* no doubt the whole world was even then in a "state of chassis" but precious little could or would those two do about it, save use it as an excuse to sustain their own weakness, incompetence, and submission to the death wishes of the society about them.[1] One wonders if something similar is going on in this nation, at this time. Having persistently failed to do what was necessary and possible to do for urban life on grounds that conditions surely were not so bad as to warrant such exertion, the nation seems suddenly to have lurched to the opposite position of declaring that things are indeed so very bad that probably nothing will work anyway. The result either way is paralysis and failure. It is time for a measure of perspective.

I take it Lewis Mumford intended to convey something of this message in his recent book *The Urban Prospect*[2] which he begins with a short preface cataloguing the ills of the modern city with a vigor and specificity that command instant assent from the reader. "Exactly!" one responds. "That is precisely the way things are! Mumford is really 'telling it like it is'." (A measure of *negritude* having of late become the mark of an authentic urban crisis watcher.) One reads on with increasing recognition, mounting umbrage, only to find at the end that this foreword was in fact written for the July, 1925, edition of *Survey Graphic*. Things have changed, but not that much, and in directions that were, in a sense, fully visible to the sensitive eye nearly a half century ago. To be sure, at a certain point, a matter of imbalance becomes one of pathology, a tendency becomes a condition, and for societies as for individuals, there comes a point when mistakes are no longer to be undone, transgressions no longer to be forgiven. But it is nowhere clear that we have reached such a point in our cities.

Continuity and change: these are the themes of all life, and not less that of cities. However, as in so many aspects of our national experience, Americans seem more aware of and more sensitive to modes of change than to those of continuity. This is surely a survival from the frontier experience. There has not, I believe, ever been anything to match the rapidity, nay fury, with which Americans set about founding cities in the course of the seventeenth, eighteenth, and nineteenth centuries. Only just now is the historical profession beginning to catch up with that breathless undertaking. Before long we are likely to have a much clearer idea than we now do as to how it all began. But it is nonetheless possible at this early state, as it were, to identify a half dozen or so persistent themes in the American urban experience, which nonetheless seem to evolve from earlier to later stages in a process that some would call growth, and others decay, but in a manner that nonetheless constitutes change.

## THEMES IN URBAN EXPERIENCE

### Violence

Through history—the history, that is, of Europe and Asia and that great bridge area in between—cities have been nominally, at least, places of refuge, while the countryside has been the scene of insecurity and exposure to misfortune and wrongdoing. Obviously the facts permit no generalization, but there is at least a conceptual validity, a persistence over time, of the association of the city with security. In the classical and feudal world, to be outside the gates was to be in trouble. Writing of the destruction of Hiroshima and Nagasaki, George Steiner evokes the ancient certainty of this point, and suggests the ways in which it lives on.

> In those two cities, the consequences have been more drastic and more specialized. Therein lies the singularity of the two Japanese communities, but also their symbolic link with a number of other cities in history and with the role such cities have played in man's consciousness of his own vulnerable condition—with Sodom and Gomorrah, visited by such fiery ruin that their very location is in doubt; with Nineveh, raked from the earth; with Rotterdam and Coventry; with Dresden, where in 1944, air raids deliberately kindled the largest, hottest pyre known to man. Already, in the "Iliad," the destruction of a city was felt to be an act of peculiar finality, a misfortune that threatens the roots of man. His city smashed, man reverts to the unhoused, wandering circumstance of the beast from which he has so uncertainly emerged. Hence the necessary presence of the gods when a city is built, the mysterious music and ceremony that often attend the elevation of its walls. When Jerusalem was laid waste, says the Haggada, God Himself wept with her.[3]

Little of this dread is to be encountered in the United States, a society marked by the near absence of internal warfare once the initial Indian conflicts were over. Warfare, that is to say, between armies. We have, on the other hand, been replete with conflict between different groups within the population, classified in terms of race, class, ethnicity, or whatever, and this conflict has occurred in our cities, which in consequence have been violent places.

An account of the draft riots in New York City in 1863 strikes a surpassingly contemporary note:

> Nothing that we could say, could add to the impressiveness of the lesson furnished by the events of the past year, as to the needs and

dangerous condition of the neglected classes in our city. Those terrible days in July—the sudden appearance, as if from the bosom of the earth, of a most infuriated and degraded mob; the helplessness of property-holders and the better classes; the boom of cannon and rattle of musketry in our streets; the skies lurid with conflagrations; the inconceivable barbarity and ferocity of the crowd toward an unfortunate and helpless race; the immense destruction of property —were the first dreadful revelations to many of our people of the existence among us of a great, ignorant, irresponsible class, who were growing up here without any permanent interest in the welfare of the community or the success of the Government—the *proletaires* of the European capitals. Of the gradual formation of this class, and the dangers to be feared from it, the agents of this Society have incessantly warned the public for the past eleven years.[4]

*—Eleventh Annual Report*
*Children's Aid Society, New York*

In some degree this violence—or the perception of it—seems to have diminished in the course of the 1930's and 1940's. Urban scholar James Q. Wilson has noted the stages by which, for example, the treatment of violence as an element in American politics steadily decreased in successive editions of *Politics, Parties, and Pressure Groups*, V.O. Key's textbook on American politics that appeared during the latter part of this period. It may be that Depression at home and then war abroad combined to restrict opportunity or impulse. But in any event, there was something of a lull, and in consequence all the more alarm when violence reappeared in the mid-1960's. But it was only that: a reappearance, not a beginning.

Yet with all this it is necessary to acknowledge a transformation however subtle and tentative. The tempo of violence seems to have speeded up, the result more or less direct of change in the technology of communications, which now communicate not simply the fact, but also the spirit of violent events, and do so instantaneously. More ominously, there appears to have been a legitimation of violence, and a spread of its ethos to levels of society who have traditionally seen themselves, and have been, the respositories of stability and respect for, insistence upon, due process. It is one thing to loot clothing stores—Brooks Brothers was hit in 1863—to fight with the police, to seize sections of the city, and hold out against them. It is another thing to seize university libraries, and that is very much part of the violence of our time, a violence that not only arises among the poor and disinherited, but also among the well to do and privileged, with the special fact that those elements in society which normally set standards of conduct for the society as a whole have been peculiarly unwilling, even unable, to protest the massive disorders of recent times.

## Migration

The American urban experience has been singular in the degree to which our cities, especially those of the North and East, have been inundated by successive waves of what might be called rural proletarians, a dispossessed peasantry moving—driven from other people's land in the country—to other people's tenements in the city. American cities have ever been filled with unfamiliar people, acting in unfamiliar ways, at once terrified and threatening. The great waves of Catholic Irish of the early nineteenth century began the modern phase of this process, and it has never entirely stopped, not so much culminating as manifesting itself at this time in the immense folk migration of the landless Southern Negro to the Northern slum. In small doses such migrations would probably have been easily enough absorbed, but the sheer mass of the successive migrations has been such as to have dominated the life of the cities in their immediate aftermath. The most dramatic consequence was that popular government became immigrant government: in the course of the nineteenth century, great cities in America came to be ruled by men of the people, an event essentially without precedent in world history. And one typically deplored by those displaced from power in the course of the transformation. Let me cite to you for example, a schoolboy exercise written in 1925, by a young Brahmin, the bearer of one of Boston's great names, on the theme "That there is no more sordid profession in the world than *Politics.*"

The United States is one of the sad examples of the present form of government called democracy. We must first remember that America is made up of ignorant, uninterested, masses, of foreign people who follow the saying, "that the sheep are many but the shepards are few." And the shepards of our government are wolves in sheeps clothing. From Lincolns Gettysburg address let me quote the familiar lines " a government of the people, for the people, and by the people." In the following lines I shall try and show you how much this is carried out in modern times.

Let us take for example the position of our mayors. They are elected by majority vote from the population in which they live. Let us take for a case Mayor Curley of Boston. He tells the Irish who make up the people of Boston that he will lower their taxes, he will make Boston the greatest city in America. He is elected by the Irish mainly because he is an Irishman. He is a remarkable politician he surrounds himself by Irishmen, he bribes the Chief Justice of the court, and although we know that the taxes that we pay all find a way into his own pocket we cannot prove by justice that he is not a just and good mayor.

But such distaste was not wholly groundless. The migrant peasants did and do misbehave: as much by the standards of the countryside they leave behind, as by those of the urban world to which they come. The process of adapting to the city has involved great dislocations in personality and manners as well as in abode. From the first, the process we call urbanization, with no greater specificity than the ancient medical diagnosis of "bellyache" or "back pain," has involved a fairly high order of personal and social disorganization, almost always manifesting itself most visibly in breakdown of social controls, beginning with the most fundamental of controls, those of family life. The Children's Aid Society of New York was founded in response to the appearance of this phenomenon among the immigrant Irish. Let me quote from their first Annual Report:

> It should be remembered that there are no dangers to the value of property, or to the permanency of our institutions, so great as those from the existence of such a class of vagabond, ignorant, and ungoverned children. This "dangerous class" has not begun to show itself as it will in eight or ten years, when these boys and girls are ma-

tured. Those who were too negligent or too selfish to notice them as children, will be fully aware of them as men. They will vote. They will have the same rights as we ourselves, though they have grown up ignorant of moral principle, as any savage or Indian. They will poison society. They will perhaps be embittered at the wealth and the luxuries they never share. *Then let society beware, when the outcast, vicious, reckless multitude of New York boys, swarming now in every foul alley and low street, come to know their power and* use *it!*[5]

Lewis Mumford speaks of precisely the same phenomenon:

> One of the most sinister features of the recent urban riots has been the presence of roaming bands of children, armed with bottles and stones, taunting and defying the police, smashing windows and looting stores. But this was only an intensification of the window-breakings, knifings, and murders that have for the past twenty years characterized "the spirit of youth in the city streets.[6]

And note the continuity of his last phrase, which alludes, of course, to Jane Addams' book, *The Spirit of Youth and the City Streets,* in which she describes just those conditions at the turn of the century in terms that William James declared "immortal" and which, we must allow, were hardly ephemeral.

Yet here too technology seems to have been playing us tricks, accentuating and exacerbating our recent experience. The newest migrants come upon an urban world that seems somehow to need them less, to find them even more disturbing and threatening, and to provide them even less secure a place in the scheme of things than was ever quite the case with those who preceded them. I take this to be almost wholly a function of changing employment patterns consequent upon changing technology. But this very technology has also provided an abundance of material resources—and a measure of social conscience—so that people who are not especially needed are nonetheless provided for, so that for example, after seven years of unbroken economic expansion, by 1968 there were 800,000 persons living on welfare in New York City, with the number expected to exceed one million in 1969. In part this is a phenomenon of birth rates.

One person in ten, but one baby in six, today is Negro. The poor continue to get children, but those children no longer succumb to cholera, influenza, and tuberculosis. Thus progress more and more forces us to live with the consequences of social injustice. In a more brutal age the evidence soon disappeared!

## Wealth

Those who have moved to cities have almost invariably improved their standard of life in the not-very-long run. Nor has this been wholly a matter of the consumption of goods and services. "City air makes men free," goes the medieval saying, and this has not been less true for industrial America. The matter was settled, really, in an exchange between Hennessy and Dooley at the turn of the century. The country, said that faithful if not always perceptive patron, is where the good things in life come from. To which the master responded, "Yes, but it is the city that they go to." Technology is at the base of this process. The standard of life in American cities rises steadily, and there are few persons who do not somehow benefit. And yet this same technology—wealth—takes its toll. More and more we are conscious of the price paid for affluence in the form of man-made disease, uglification, and the second and third order effects of innovations which seem to cancel out the initial benefits.

Nathan Keyfitz has nicely evoked the paradox implicit in many of the benefits of technology. Plenty encourages freedom. It also encourages density. Density can only be managed by regulation. Regulation discourages freedom. The experienced, conditioned city dweller learns, of course, to live with density by maintaining, as Keyfitz puts it, "those standards of reserve, discretion, and respect for the rights of others" that keep the nervous system from exhausting itself on the overstimulus available on any city street.[7] The traditional assertion of Manhattan apartment dwellers that they have never met their neighbors across the hall is not a sign of social pathology: to the contrary, it is the exercise of exemplary habits of social hygiene. Borrowing a meter from Canning's account of the failings of the Dutch, the rule for the modern cliff dweller might be put as follows:

In the matter of neighbors,
The sound thing to do,
Is nodding to many
But speaking to few.

It may be speculated, for example, that a clue to the transformation of the roistering, brawling, merrie England of tradition into that somber land where strangers dare not speak to
one another in trains lies *in the fact of the trains.* Technology—
in this case the steam engine which created the vast nineteenth
century complexes of London and Manchester—brought about
urban densities which required new forms of behavior for those
who would take advantage of or at very least survive them. The
British, having been first to create the densities, were first to
exhibit the telltale *sangfroid* of the modern urban dweller.

It may also be speculated that the "disorganized" life of rural
immigrants arises in some measure at least from an inability to
control the level of stimulus: to turn down the radio, turn off
the television, come in off the streets, stay out of the saloons,
worry less about changing styles of clothes, music, dance, whatever. Lee Rainwater has provided us with painful accounts of
the feeling of helplessness of the mothers of poor urban families
in the face of the incursions from the street: the feeling, literally,
that one could not simply close one's door in the housing project,
and refuse to allow family, friends, neighbors, and God knows
whom else to come and go at will. This makes for lively neighborhoods, which from a distance seem almost enviable, but also
for very disturbed people.

When such groups become large enough, when densities become ominous, government regulation becomes necessary or, at
least, all but invariably makes its appearance, so that even for
the disciplined urbanite, technology at some point begins to
diminish freedom. Keyfitz writes:

George Orwell's *1984* is inconceivable without high population
density, supplemented by closed circuit television and other devices
to eliminate privacy. It exhibits in extreme form an historical process by which the State has been extending its power at the expense

of the Church, the Family, and the Local Community, a process extending over 150 years.[8]

There are but few bargains in life, especially in city life.

## Mobility

Cities are not only places where the standards of life improve, but also very much—and as much now as ever—they are places where men rise in social standing. Otis Dudley Duncan and Peter Blau in their powerful study, *The American Occupational Structure*,[9] have made this abundantly clear. American cities are places where men improve their position, as well as their condition or, at least, have every expectation that their sons will do so. The rigidities of caste and class dissolve, and opportunity opens. Yet this has never been quite so universally agreeable an experience as one could be led to suppose from the accounts of those for whom it has worked. If in the city men first, perhaps, come to know success, it is there also that often, especially those from the most caste-ridden rural societies, they first come to know failure. It seems to me that this is a neglected aspect of the urban experience. I would argue that the rural peasant life of, let us say, the Irish, the Poles, the Slavs, the Italians, and the Negro Americans who have migrated over the past century and a half was characterized by a near total absence of opportunity to improve one's position in the social strata, but also it was characterized by the near impossibility of observing others improve theirs. Rarely, in either absolute or relative terms, did individuals or families of the lowest peasant classes experience decline and failure: that in a sense is the law of a noncontingent society. Only with arrival in the city does that happen, and I would argue that for those who lose out in that competition, the experience can be far more embittering than that brought on by the drab constancy of country life.

Again technology—again television, for that matter—plays its part. Stephan Thernstrom has noted that the immigrant workers of nineteenth-century New England, earning $1.50 a day when they had work, nonetheless managed in surprising numbers to put aside some money and to buy a piece of property and re-spectability before their lives were out, despite the fact that their incomes rarely permitted them to maintain the minimum stand-ard of living calculated by the social workers of the time.[10] The difference, Thernstrom notes, was that for the migrants a mini-mum standard of living was potatoes. Period. So long as they did not share the expectations of those about them—even the small expectations of social workers—they were not deprived. But advertising and television, and a dozen similar phenomena, have long since broken down that isolation, and the poor and newly arrived of the American city today appear to be caught up in a near frenzy of consumer emotions: untouched by the disenchantment with consumption of those very well off, and unrestrained by the discipline of household budgets imposed on those in between. The result, as best such matters can be measured, is a mounting level of discontent, which seems to slide over from the area of consumption as such to discontent with levels of social status that do not provide for maximum levels of consumption, so that even those who might be thought to be succeeding in the new urban world appear to feel they are not succeeding enough, while others are suffused with a sense of failure.

### Intellectual Disdain

". . . Enthusiasm," Morton and Lucia White write, "for the American city has not been typical or predominant in our in-tellectual history. Fear has been the more common reaction."[11] Fear, distaste, animosity, ambivalence. "In the beginning was the farm," or so the Jeffersonian creed asserts. And the great symbol—or perhaps consummation would be the better term—of this belief was the agreement whereby in return for the Jeffersonian willingness to have the federal government accept the debts acquired by states during the Revolutionary War, the capital of the new nation would be transferred from the City of New York to a swamp on the banks of the Potomac. Do not suppose that that agreement has not affected American history. New York remains the capital of the nation, as that term is usually understood, in the sense of the first city of the land. It is the capital of finance, art, theater, publishing, fashion, intellect, industry . . . name any serious human endeavor other than politics, and its center in the United States will be found in New York City. In years of hard-fought presidential primaries, it is even for many purposes the political capital of the nation. But the seat of government is in Washington, which is only just beginning to respond to the fact that for half a century now ours has been a predominantly urban society.

Once again technology seems to be interacting with a pre-existing tendency. As the American city came more and more to be the abode of the machine, the alarm of American intellectuals, if anything, was enhanced. And to a very considerable degree legitimated, for surely machines have given a measure of reality to alarums that were previously more fantasy than otherwise. To this has been added an ever more persistent concern for social justice, so that American intellectuals now conclude their expanding catalogues of the horrors of urban life with ringing assertions that the cities must be saved. But it is to be noted that this comes almost as an afterthought: the conviction that in the cities will be found the paramount threat to the life of the Republic has changed hardly at all. What has perhaps changed is that at long last what they have been saying may be beginning to be true.

## Ugliness

That there are great and stunning exceptions is as much a matter of accident as anything. The essential fact is that for all the efforts to sustain and assert a measure of elite concern for urban aesthetics—of the kind one associates with historical preservation societies—and for all the occasional bursts of energy within the urban planning profession, the American city remains an ugly place to live, pretty much because we like it that way. A measure, no doubt, of this persisting condition can be attributed to the business and propertied interests of the nation that have resisted municipal expenditure, notably when it passed through the hands of egalitarian city halls. But it is more than that.

Somehow, somewhere in the course of the development of democratic or demagogic tradition in this nation, the idea arose that
concern for the physical beauty of the public buildings and
spaces of the city was the mark of—what?—cryptodeviationist
anti-people monumentalism—and in any event an augury of de-
feat at the polls. The result has been a steady deterioration in
the quality of public buildings and spaces, and with it a decline
in the symbols of public unity and common purpose with which
the citizen can identify, of which he can be proud, and by
which he can know what he shares with his fellow citizens. For
the past seven years, as an example, I have been involved with
efforts to reconstruct the center of the city of Washington, an
attempt that begins with the assertion of the validity and viability of L'Enfant's plan of the late eighteenth century. In this
effort we have had the tolerant to grudging cooperation of a
fairly wide range of public and private persons, but let me say
that we have had at no time the enthusiasm of any. And now
I fear we may have even less, since of late there has arisen the
further belief that to expend resources on public amenities is in
effect to divert them from needed areas of public welfare. The
very persons who will be the first to demand increased expenditures for one or another form of social welfare will be the
last to concede that the common good requires an uncommon
standard of taste and expenditure for the physical appointments
of government and the public places of the city.

This attitude was perhaps unintentionally evoked by the respected Episcopal Bishop of New York who in 1967 announced
that, in view of the circumstances of the poor of the city, he
would not proceed with the completion of the Cathedral of St.
John the Divine, the largest such building ever begun, situated
on a magnificent site overlooking the flat expanse of Harlem.
Why? Meaning no disrespect, is it the plan of the church to
liquidate its assets and turn them over to the poor? How much
would that come to per head? But even so, would not the completed cathedral be an asset? If men need work, could they not
be given jobs in its construction? The French—*toujours gai,* as
Mehitabel would have it—built Sacré-Coeur as an act of penance
for the excesses of the Commune. Could not the Episcopalians

build St. John the Divine—a perfect symbol of rebirth—as a gesture of penance for all that brahmin disdain which, in one form or another, to use Max Ways' phrase, taught us to despise our cities until they become despicable.

If the phenomenon of ugliness, the last of my urban themes, can be thought to have arisen from more or less abstract qualities of American society, in the present and foreseeable future its principal cause is visible, concrete, and ubiquitous, which is to say it is the automobile. More than any other single factor it is the automobile that has wrecked the twentieth-century American city, dissipating its strength, destroying its form, fragmenting its life. So utterly pervasive is the influence of the automobile that it is possible almost not to notice it at all. Indeed, it is almost out of fashion to do so: the men who first sought to warn us have almost ceased trying, while those who might have followed have sought instead formulations of their own, and in that manner diverted attention from the essential fact in the age of the automobile; namely, that cities, which had been places for coming together, have increasingly become machines for moving apart, devices whereby men are increasingly insulated and isolated one from the other.

## TECHNOLOGY AND THE CITY

### The Automobile

A coda of sorts that has persisted through the elaboration of the successive themes of this chapter has been the recent role of technology in accentuating and in a sense exacerbating long-established tendencies. The impact of technology on human society—on all forms of life—is the pre-eminent experience of the modern age, and obviously of the city as well. But only of late, one feels, has any very considerable appreciation developed that, after a point, change in quantity becomes change in quality, so that a society that begins by using technology can end by being used by it, and in the process, somehow, loses such control of its destiny as past human societies can be said to have done. The sheer rationality of technology has argued against any such assumption which has seemed more allied to a lack of comprehension than otherwise. People fear what they do not understand, and the command of technological skills among critics of technology has tended to be minimal, hence the presumption that warnings of disaster, while interesting, were at the same time self-serving.

One begins to think that this may not be so. Take the family automobile. A simple, easily enough comprehended (or seemingly so), unthreatening, and convenient product, if anything, of folk technology rather than of modern science. Who would imagine any great harm coming from the automobile? Yet consider a moment. With the advent of the automobile everyday citizens, for the first time in human history, came into possession of unexampled physical energy: the powers of the gods themselves became commonplace. And from the very outset violence ensued. In 1895, for example, when there were only four gasoline-powered vehicles in the United States, two were in St. Louis, Missouri, and managed to collide with such impact as to injure both drivers, one seriously. Thus was introduced a form of pathology that was to grow steadily from that year to this. Today, somewhere between one-quarter and two-thirds of the automobiles manufactured in the United States end up with blood on them. Indeed so commonplace and predictable have collisions become that the U.S. Court of Appeals for the Eighth Circuit recently ruled in *Larsen v. General Motors* that a crash must be considered among the "intended uses" of a motor vehicle, and the manufacturers accordingly responsible to provide for such contingency in their design.[12]

It becomes increasingly clear that the major environment, or, if you will, vehicle, in which incidents of uncontrolled episodic violence occur within the population is that of the automobile. Whether access and exposure to this environment has increased the incidence of such episodes, or whether the urban environment now largely created and shaped by the automobile has generally increased the violence level is an uncertain question at best, (there has, of course, been a great decline in violence directed toward animals) but with the number of deaths and injuries at the present ongoing rates, and the number of vehicles in use approaching the 100 million mark, it is a matter worth pondering.

Crashes are but one form of pathology. Each year in the United States, automobiles pour 86 million tons of carbon monoxide, oxides of nitrogen, hydrocarbons, sulfur oxides, lead compounds, and particulates into the air we breathe. So much that recently my youngest son came home with a button that announced "Clean air smells funny." Dr. Claire C. Patterson of the California Institute of Technology put it another way in testimony before a Congressional committee: "The average resident of the United States is being subject to severe chronic lead insult," originating in lead tetraethyl. Such poisoning can lead to severe intellectual disability in children; so much that Patterson feels it is dangerous for youth to live for long periods of time near freeways.

But that is only the beginning, hardly the end of the impact of this particular form of technology on the society at this time. In consequence of the management of the automobile traffic system by means of traditional rules of the road, the incidence of armed arrest of American citizens is the highest of any civilization in recorded history. In 1965, for example, the California Highway Patrol alone made one million arrests. Indeed so commonplace has the experience become that a misdemeanor or felony committed in a motor vehicle is no longer considered a transgression of any particular consequence, and to be arrested by an armed police officer is regarded as a commonplace. That is precisely what Orwell told us would happen, is it not?

There are some 13.6 million accidents a year, with some 30 million citations for violations issued each twelve months. And at this point, ineluctably, technology begins to impact on the most fundamental of civil institutions, the legal system itself. Largely in consequence of the impact of traffic crash litigation, it now takes an average of 32.4 months to obtain a civil jury trial for a personal injury case in the metropolitan areas of the nation. In Suffolk County, New York, it is 50 months. In the Circuit Court of Cook County, serving Chicago, it is 64 months. This past winter in Bronx County, New York, the Presiding

Judge of the Appellate Division announced he was suspending civil trials altogether while he tried to catch up with criminal cases. The courts are inundated, the bar is caught up, implicated and confused, the public knows simply that somehow justice is delayed and delayed. All of which is a consequence of this simplest form of technology, working its way on the institutions of an essentially pre-technological society.

## Technology as Art

It sometimes happens that a work of art appears at a particular moment in time and somehow simultaneously epitomizes and reveals the essential truths of the time. In a most astonishing and compelling way this has happened for the American city, and it has done so, most appropriately, on Forty-second Street in Manhattan, in the persona, if you will, of Kevin Roche's Ford Foundation headquarters. So much comes together here. A great firm of architects, successor to Eero Saarinen Associates whose first large commission was, of course, the General Motors Technical Center outside Detroit, Saarinen, and now Roche, have gathered a group of artist-technicians whose work, from the Dulles Airport at Chantilly, Virginia, to the Trans World Airlines Terminal at Kennedy Airport and the Columbia Broadcasting headquarters in New York City, has evoked the power and purpose of the age of technology as perhaps no other organization has. Aline B. Loucheim once wrote of her future husband Eero Saarinen, that his contribution lay in "giving form or visual order to the industrial civilization to which he belongs, designing imaginatively and soundly within the new esthetics which the machine age demands and allows." Allan Temko describes the General Motors buildings as a "fusion of serious design and technological power". *Architectural Forum* called them "exalted industrial products."

Here in the Ford Foundation headquarters is expressed the very highest purposes of that technological power: compassionate and potent concern for the betterment of man's lot. The building is everything a building could be: a splendid workplace, a gift to the city in the form of a public park, a gift to the world simply as a work of imagination and daring. If it be a reproach of sorts to the public and private builders of the nation who by and large show little of either, it is a gentle reproach, more show than tell. In that favored form of foundation giving, it is a demonstration project of sorts: an example of what can, and what therefore, in the formulation of Jacques Ellul, in an age of technology must be done.[13]

The exterior of the building is quiet and unassertive: it is not a big building, and seeks rather to understate both its size and importance. Shafts of Corton steel rise from the ground, here and there sheathed with a blue-brown granite, and interspersed with large rectangular glass panels. Rather in the mode of a cathedral, the portals do not so much impart as suggest the experience to come. It is only on entering—Chartres, say, or Vezalay —and encountering the incomparable space, shaped and reserved for a single purpose only, that one leaves off observing the building and begins to be shaped by it: the eye rises, the mind turns to last things. So with the Ford Foundation headquarters. One passes through revolving doors to enter a garden. Truly, a garden, a small park, like nothing anywhere else to be encountered, a third of an acre, lush and generous, climbing a small hill that follows the terrain of Manhattan at this point, illuminated by the now vast windows that climb nine stories towards heaven itself, and there only to be met by a glass roof. Water moves slightly in a pool—a font? Attendants move quietly, and are helpful. One notices that vegetation sprouts from beams and ledges on the third and fourth and even the fifth floors. One is awestruck by the wealth and power of the Foundation, and the sheer authority of its intent to do good. Only the gray-white light is not quite what it should be: as in those French and German cathedrals whose stained glass was lost to war, or revolution, or Protestantism.

But this is only the entering light. As in any such edifice, there is a light within. In this case a very monstrance-like golden brown glow that shines forth from the offices of the Foundation executives, who from the floor of the park are to be seen at their work behind glass panels formed and reticulated by the same rusted beams that frame the colorless glass of two sides of the building. (Corton steel seals itself by rusting and need not be painted.) At this point one perceives readily enough that the building has been built as a factory. Not precisely as a factory—any more than the Gothic revival built office buildings precisely as medieval monasteries—but rather to evoke the style and somehow the spirit of a great plant. The huge, heavy lateral beams, from which elsewhere would be suspended the giant hoists that roam back and forth amidst the clatter and roar, the sawtooth roof, the plant managers' eyrie hung from the ceiling, keeping an eye on everything, the perfectly standardized, interchangeable fixtures in each office, the seriousness and competence of it all, even the blue-black, somehow oily granite of the cheerless rest rooms ("No Loitering in the Can") magically, stunningly, triumphantly, evoke the style and spirit of the primeval capitalist factory. Corton. Red. Rouge. River Rouge. Of course! And why not for $16 million of Henry Ford's money? He was that kind of man. Knew how to make automobiles and obviously liked to. Else he could hardly have done it so well. All black, just as the Ford Foundation headquarters is all brown. Same principle. So also the Panopticon effect of the exposed offices wherein the presumptively interchangeable officers at their perfectly interchangeable desks labor at their good works in full view of management and public alike. (The public serving perhaps as the visitors to Bentham's prospective model prison: a "promiscuous assemblage of unknown and therefore unpaid, ungarbled and incorruptible inspectors"?) Critics, at least in the first reviews, seem to have missed most of this, but no matter: the architecture needs no guide book: the intellectual and aesthetic effect is not to be avoided, even when the intent is least perceived. All in all it is just as McGeorge Bundy proclaimed it in the 1968 annual report of the Foundation: "Kevin Roche's triumph."

But it is more than that. Or rather there is more than is to be perceived at one time. A great work of art has levels of meaning at once various and varying. Standing in the park, gazing upwards, following the factory motif, the mind is suddenly troubled. Something is missing. Noise. Factories are places of noise. Of life. Clatter. Roar. There is no noise here. Only quiet. The quiet of the. . .? The mind oscillates. It is a factory, all right. But a ruined factory! The holocaust has come and gone: hence the silence. The windows have blown out, and only the gray light of the burnt-out world enters. The weather has got in, and with it nature now reclaiming the ravaged union of fire and earth. The factory floor has already begun to turn to forest. Vegetation has made its way to ledges halfway up the interior. (Trust Dan Kiley to understand.) The machine tools are gone. Reparations? Vandalism? Who knows? But the big machines will no longer be making little machines. Gone too is the rational, reforming, not altogether disinterested purpose of the Panopticon.[14] One is alone in the ominous gloom of a Piranesi prison, noting the small bushes taking hold in the crevices of the vast ruined arches.

Is it the past or the future that has taken hold of the mind? Certainly the ruined steel frame is a good enough symbol of the twentieth century so far. (Where had one last seen that color? Of course. Pingree Street in Detroit after the riot. A year later there it was again on Fourteenth Street in Washington: the fiery orange-red of the twisted steel shopping centers' framing after the looting and arson has passed.) Or is it the future? There is *surréal* quality that comes of standing in the ruined half of the building, watching the life going on behind the glass walls of the intact half, seemingly oblivious to the devastation without. Can ruin advance slowly like rot? No. Yes. Did the automobile start all this? No. Surely it is all this that started automobiles. One-quarter to two-thirds of which end up with blood on them. Blood. Red. Rouge. River Rouge.

Enough.

But then why has Stein built the Ford Foundation head-quarters in New Delhi immediately adjacent to the Lodi Tombs, symbols of death and sensual to the point of necrophilia? Did not Bentham remark that he could legislate wisely for all India from the recesses of his study? There's a Panopticon in your future.

No. Enough.

## Architecture and "Community"

And yet it comes together in a way. *"Le siècle de la machine,"* Le Corbusier wrote in 1924, *"a réveillé l'architecte."* Not least because the machine destroys so much of that experience of community that the architect seeks to create. Allan Temko in his study of Eero Saarinen describes Saarinen's purpose in these terms. "What . . . [he] wished to renew, maintain, and improve was the organic expression of the *civitas* which he found weakened or destroyed virtually everywhere in modern civilization, with one significant exception—the university campus." And so Roche built a ruined machine-for-making-machines as the headquarters of a great philanthropic foundation whose principal concerns have been to support the universities of the nation, and to seek to strengthen the community life of its cities.

The research of James Q. Wilson and Edward C. Banfield at Harvard University is now beginning to produce results surprisingly similar to the visions of the architect-artist. A thousand Boston homeowners were asked what they thought to be the biggest problem facing the city. As Wilson puts it, "After a decade or more of being told by various leaders that what's wrong with our large cities is inadequate transportation, or declining retail sales, or poor housing, the resident of the big city is beginning to assert his own definition of that problem—and this definition has very little to do with the conventional wisdom on the urban crisis."[15] Wilson and his colleague asked one thousand Boston homeowners what they thought to be the biggest urban problem of this time.

> The "conventional" urban problems—housing, transportation, pollution, urban renewal, and the like—were a major concern of only 18 per cent of those questioned, and these were expressed disproportionally by the wealthier, better-educated respondents. Only 9 per cent mentioned jobs and employment, even though many of those interviewed had incomes at or even below what is often regarded as the poverty level. *The issue which concerned more respondents than any other was variously stated—crime, violence, rebellious youth, racial tension, public immorality, delinquency. However stated, the common theme seemed to be a concern for improper behavior in public places.*
>
> What these concerns have in common, and thus what constitutes the "urban" problem for a large percentage (perhaps a majority) of urban citizens, is *a sense of the failure of community.*[16]

And yet cities, by definition, destroy community. Or is it only when they are too big, too unsettled, that they do this? Is it only when social conditions are allowed to arise which lead inevitably to asaults on the private communities which experienced city dwellers create for themselves, which in turn lead to more collective regulation, and in consequence less of the self-imposed decision to behave properly and as expected, which is the essence of community?

We do not know. "Them what gets the apple, gets the worm," goes the Negro saying. Is that what the Ford Foundation building represents: a shining exterior, rotting from within? A civilization whose cancerous growth has already devoured half its offspring, and is moving towards the unthinking, untroubled other half? We shall see. Hopefully in the meantime we shall also think about it a bit. Mumford, unfailingly, has sorted out the levels of immediacy and difficulty of the current crisis.

> To go deeper into this immediate situation we must, I suggest, distinguish between three aspects, only one of which is open to immediate rectification. We must first separate out the problems that are soluble with the means we have at hand: this includes such immediate measures as vermin control, improved garbage collection, cheap public transportation, new schools and hospitals and health clinics. Second, those that require a new approach, new agencies, new methods, whose assemblage will require time, even though the earliest possible action is urgent. And finally there are those that require a reorientation in the purposes and ultimate ideals of our whole civilization—solutions that hinge on a change of mind, as far-reaching as that which characterized the changes from the medieval religious mind to the modern scientific mind. Ultimately, the success of the first two changes will hinge upon this larger—and, necessarily, later —transformation. So, far from looking to a scientifically oriented technology to solve our problems, we must realize that this highly sophisticated dehumanized technology itself now produces some of our most vexatious problems, including the unemployment of the unskilled.[17]

But something more than thinking will be required. A certain giving of ourselves with no certainty what will come of it. It is the only known way, and imperfectly known at that.

Attend to Mrs. Boyle at the end of *Juno and the Paycock,*
pleading for the return of a simpler life, a life before all things
had become political, before all men were committed, before
all cities somehow seemed in flames:

Sacred Heart o' Jesus take away our hearts o' stone,
and give us hearts o' flesh. Take away this murdherin'
hate, and give us Thine own eternal love!

## Notes

1. Sean O'Casey, *Juno and the Paycock* (London: MacMillan & Co., Ltd. 1950.)
2. Lewis Mumford, *The Urban Prospect* (New York: Harcourt, Brace & World, Inc., 1968.)
3. George Steiner, "White Light in August," *The New Yorker* (August 3, 1968), p. 76.
4. *Eleventh Annual Report of the Children's Aid Society of New York* (New York: Press of Wynkoop, Hallenback, and Thomas, February, 1864), p. 3.
5. *First Annual Report of the Children's Aid Society of New York* (New York: Printed by C. W. Benedict, February, 1854), p. 12.
6. Mumford, *op cit.,* p. 251.
7. Nathan Keyfitz, "Population Density and the Style of Social Life," *Bio-Science,* XVI (Washington, D.C.: American Institute of Biological Sciences, December, 1966.)
8. *Ibid.*
9. Peter M. Blau and Otis Dudly Duncan, *The American Occupational Structure* (New York: John Wiley & Sons, Inc. 1967.)
10. Stephan Thernstrom, *Poverty and Progress* (Cambridge: Harvard University Press, 1964.)
11. Morton and Lucia White, *The Intellectual Versus the City* (Cambridge: Harvard University and the M.I.T. Press), p. 1.
12. *Erling David Larsen v. General Motors.* U.S. Court of Appeals, 8th Circuit 18, 853 (March 11, 1968.) A not unrelated form of technological style is involved in this decision. The concept of the probability of crash per vehicle was developed by Haddon and me in lectures at the Maxwell Graduate School of Public Administration of Syracuse University in 1959 and 1960. It was refined and presented as a paper by Haddon and Goddard in 1960 and published in 1962. Within six years it had made its way, by specific reference, into a Federal judicial decision. The near volatile degree to which new ideas, good and bad, are put into practice in a technologically advanced society is very likely a source of the common complaint about unsettledness.

13. Jacques Ellul, *The Technological Society*, trans. John Wilkinson (New York: Alfred A. Knopf, 1964.)

14. See Gertrude Himmelfarb, "The Haunted House of Jeremy Bentham," *Victorian Minds*. (New York: Alfred A. Knopf, 1968.)

15. James Q. Wilson, "The Urban Unease: Community vs. City," *The Public Interest* (Summer, 1968), p. 25.

16. *Ibid.*, p. 26.

17. Mumford, *op. cit.*, p. 246.

# Index

345

Lewis and Clark College - Watzek Library
HT123 .T66
/Toward a national urban policy      wmain

3  5209  00414  8728